Springer Theses

Recognizing Outstanding Ph.D. Research

Aims and Scope

The series "Springer Theses" brings together a selection of the very best Ph.D. theses from around the world and across the physical sciences. Nominated and endorsed by two recognized specialists, each published volume has been selected for its scientific excellence and the high impact of its contents for the pertinent field of research. For greater accessibility to non-specialists, the published versions include an extended introduction, as well as a foreword by the student's supervisor explaining the special relevance of the work for the field. As a whole, the series will provide a valuable resource both for newcomers to the research fields described, and for other scientists seeking detailed background information on special questions. Finally, it provides an accredited documentation of the valuable contributions made by today's younger generation of scientists.

Theses are accepted into the series by invited nomination only and must fulfill all of the following criteria

- They must be written in good English.
- The topic should fall within the confines of Chemistry, Physics, Earth Sciences, Engineering and related interdisciplinary fields such as Materials, Nanoscience, Chemical Engineering, Complex Systems and Biophysics.
- The work reported in the thesis must represent a significant scientific advance.
- If the thesis includes previously published material, permission to reproduce this must be gained from the respective copyright holder.
- They must have been examined and passed during the 12 months prior to nomination.
- Each thesis should include a foreword by the supervisor outlining the significance of its content.
- The theses should have a clearly defined structure including an introduction accessible to scientists not expert in that particular field.

More information about this series at http://www.springer.com/series/8790

Hugo Campelo

FluSHELL – A Tool for Thermal Modelling and Simulation of Windings for Large Shell-Type Power Transformers

Doctoral Thesis accepted by
the University of Porto, Portugal

 Springer

Author
Dr. Hugo Campelo
Transformers R&D Department
EFACEC Energia, S.A.
Porto
Portugal

Supervisors
Prof. José Carlos Lopes
Department of Chemical Engineering
Faculty of Engineering of the University of
  Porto
Porto
Portugal

Prof. Madalena Maria Dias
Department of Chemical Engineering
Faculty of Engineering of the University of
  Porto
Porto
Portugal

ISSN 2190-5053   ISSN 2190-5061   (electronic)
Springer Theses
ISBN 978-3-319-89199-6   ISBN 978-3-319-72703-5   (eBook)
https://doi.org/10.1007/978-3-319-72703-5

Printed on acid-free paper

This Springer imprint is published by Springer Nature
The registered company is Springer International Publishing AG
The registered company address is: Gewerbestrasse 11, 6330 Cham, Switzerland

The only true wisdom is in knowing you know nothing.

Socrates

Supervisors' Foreword

This thesis addresses a novel application of network modelling methodologies to power transformers. Network modelling is used to develop a tool to simulate the thermal performance of these machines, widely acknowledged to be critical assets in electrical networks.

After strong deregulation of electricity markets and decarbonization of worldwide economies, electrical networks have been changing fast. Both asset owners and equipment manufacturers are being driven to develop increasingly accurate simulation capabilities to optimize either their operation or their design. Temperature is a critical parameter in every electric machine, and power transformers are not an exception.

In this work, a novel thermal model has been developed and its simulation results verified against predictions of a commercial CFD code as well as experiments conducted in a dedicated set-up built exclusively for this purpose.

Hence, this work cross-links three of the most important aspects in high-quality research: model development, simulation and experimental validation. Its content is relevant to a plurality of stakeholders, from utilities to power transformer manufacturers and science community in general.

This work was funded by a Portuguese company, EFACEC Energia, one of the world leaders in power transformer technology and represents a major milestone in a long collaboration between EFACEC and FEUP, the Engineering School of University of Porto. Within this collaboration, further work has been started, namely on the development of dynamic thermal network models.

<table>
<tr><td>Porto, Portugal
June 2017</td><td style="text-align:right">Prof. José Carlos Lopes
Prof. Madalena Maria Dias</td></tr>
</table>

Abstract

The current design cycle of power transformers, in general, and shell-type transformers, in particular, demands contradicting features from the design tools. On the one hand, it demands faster responses, but on the other hand, it requires more detailed information to enable optimized decisions.

At the design stage, the thermal performance of the windings is a key characteristic to be addressed. The thermal design tools currently used are targeted to determine just the average and maximum temperatures of the windings based on a reduced number of parameters and empirical factors. Although useful and valid, these tools reflect the current design practices and do not provide means for differentiation with innovative technological solutions. Therefore, the capability of accurately predicting the detailed spatial distribution of the winding temperatures and cooling fluid velocities can be a relevant competitive advantage.

In this work, and to bridge this gap, a novel thermal-hydraulic network simulation tool has been first developed for shell-type windings—the FluSHELL tool. Its comparison against simulations on a commercial Computational Fluid Dynamics (CFD) code reveals equivalent degrees of accuracy and detail. FluSHELL shows average accuracies of 1.8 °C and 2.4 °C for the average and maximum temperatures, respectively, and the locations of the maximum winding temperatures have been consistently well predicted. The fluid mass flow rate and pressure distributions show similar trends, and both can be predicted with average deviations of 20%. Similar to CFD, this has been accomplished by discretizing the calculation domain into sets of smaller interconnected elements, but FluSHELL is observed to be approximately 100 times faster than a comparable CFD simulation.

An experimental set-up has been designed, constructed and used to prove this concept. The set-up represents the closed cooling loop of a shell-type winding, and due to its operation under DC conditions, it provides means to complement the measurements of local temperatures with accurate measurements of the average temperatures. The experimental validation showed predictions with the same trends and with average accuracies in the same order of magnitude of the combined uncertainties associated with the measurements.

Based on these results, the FluSHELL tool developed and its associated methodology are both considered conceptually validated. Further applications of this tool to commercial transformers can now be envisaged.

List of Publications

Parts of this thesis have been published in the following journal articles/conference proceedings:

H. M. R. Campelo, R. T. Oliveira, Carlos M. Fonte, X. M. López-Fernandez, M. M. Dias, José Carlos B. Lopes, "Modelling the Hydrodynamics of Cooling Channels inside Shell-Type Power Transformers with CFD.", 12th International Chemical and Biological Engineering Conference, Porto, Portugal, 2014.

H. M. R. Campelo, L. F. Braña, X. López-Fernandez, "Thermal Hydraulic Network Modelling Performance in Real Core Type Power Transformers.", 21th International Conference on Electrical Machines, Berlin, Germany, 2014.

H. M. R. Campelo, R. T. Oliveira, Carlos M. Fonte, M. M. Dias, José Carlos B. Lopes, "Modelling the Hydrodynamics of Cooling Channels inside Shell-Type Power Transformers with CFD", 3rd International Colloquium on Transformer Research and Asset Management, Split, Croatia, 2014.

H. M. R. Campelo, J. P. B. Baltazar, R. T. Oliveira, Carlos M. Fonte, M. M. Dias, José Carlos B. Lopes, "Extracting Relevant Transport Properties Using CFD Simulations of Shell-Type Electric Transformers.", ICHMT International Symposium on Computational Heat Transfer, New Jersey, USA, 2015.

H. M. R. Campelo, J. P. B. Baltazar, C. M. M. Carvalho, R. C. Lopes, R. T. Oliveira, Carlos M. Fonte, M. M. Dias, José Carlos B. Lopes, "SmarTHER Shell-Type Transformers: Integrating advanced thermal modelling techniques in the design-cycle.", 5th European Conference on HV & MV Substation Equipment, Lyon, France, 2015.

H. M. R. Campelo, J. P. B. Baltazar, C. M. M. Carvalho, R. C. Lopes, R. T. Oliveira, Carlos M. Fonte, M. M. Dias, José Carlos B. Lopes, "Novel Thermal-Hydraulic Network Model for Shell-Type Windings. Comparison with CFD and Experiments.", Cigré Session 46, Paris, France, 2016.

H. M. R. Campelo, M. A. Quintela, J. P. B. Baltazar, R. C. Lopes, C. M. M. Carvalho, "Practical Relevance of Advanced Thermal Modelling Techniques for the Modern Design and Management of Power Transformers", EuroTechCon - Primary Asset Life Management, UK, 2016.

H. M. R. Campelo, J. P. B. Baltazar, R. T. Oliveira, M. M. Dias, José Carlos B. Lopes, Carlos M. Fonte, "FLUSHELL – A Tool for Thermal Modelling and Simulation of Windings for Large Shell-Type Power Transformers", XVII ERIAC DECIMOSÉPTIMO ENCUENTRO REGIONAL IBEROAMERICANO DE CIGRÉ, Paraguay, 2017.

Acknowledgements

This journey has been long, fruitful and possible due to a significant number of high-quality persons and organizations that made part of it. In a first instance, I would like to thank my both supervisors Prof. José Carlos Brito Lopes and Prof. Madalena Dias with whom I have been working for many years and with whom I have acquired most of my competencies.

Afterwards, I would like to thank collectively EFACEC Energia for fully supporting these activities. EFACEC has always assumed the creation of knowledge as a crucial paradigm for its technological leadership. There is real and responsible research going on every day, and I sincerely hope that the market can recognize that. A significant group of colleagues and departments have been directly and indirectly involved in this work, but I would like to express my gratitude particularly to Mr. Duarte Couto and Mr. Jácomo Ramos that have always believed in me and inspired me every day. A special mention to Mr. Ricardo Lopes which is a deep transformer expert that shared his knowledge and shortened significantly the time needed to understand this machine and another special word to Mr. Carlos Carvalho who embraced this work with crucial insights into improvements in the experimental set-up.

As member of the R&D Transformers Department, Porto, I had the opportunity to witness important organizational changes along these years. Some of them more pacific than the others, as supposed, but there are two persons with whom I frequently brainstormed about how to better manage and conduct research activities inside corporate environments. They are Prof. Xose Lopez-Fernandez and Mrs. Acília Coelho.

As part of the work has been in collaboration with the University of Porto, namely its LSRE-LCM Associated Laboratory, I would also like to mention Dr. Carlos Fonte and Mr. Rómulo Oliveira who have always shown a great commitment and enthusiasm that has been reflected in significant contributions namely on the CFD part.

In addition, one of the most relevant contributions I would like to acknowledge is from Mr. José Baltazar. I had the opportunity to supervise him during his master thesis and during his internship at EFACEC. He is a highly talented and bright engineer that helped me developing this tool and participated throughout the construction and use of the experimental set-up.

At the end, I would also like to issue a collective word to all my colleagues and friends that made part of the CIGRE Working Group A2.38 and that created a unique collaborative environment. Some of these results also reflect the innumerous discussions we had together. I hope you have all enjoyed as much as I did and wish you all the best.

Contents

List of Figures

List of Tables

Notation

DP_v	Viscometric degree of polymerization [-]
g	Average winding gradient [°C]
H	Hot-spot factor [-]
Q	Factor Q [-]
S	Factor S [-]
$u_{ch,SD}$	Average oil velocity in a scaled-down fluid channel [cm·s^{-1}]
$u_{ch,FS}$	Average oil velocity in a full-scale fluid channel [cm·s^{-1}]
$Re_{ch,SD}$	Reynolds number in a scaled-down fluid channel [-]
$Re_{ch,FS}$	Reynolds number in a full-scale fluid channel [-]
$d_{h,ch,SD}$	Hydraulic diameter of a scaled-down fluid channel [m]
$d_{h,ch,FS}$	Hydraulic diameter of a full-scale fluid channel [m]
$q^v_{ch,SD}$	Volumetric flow rate in a scaled-down fluid channel [m^3·s^{-1}]
$q^v_{ch,FS}$	Volumetric flow rate in a full-scale fluid channel [m^3·s^{-1}]
$A_{f,ch,SD}$	Average flow area of a scaled-down fluid channel [m^2]
$A_{f,ch,FS}$	Average flow area of a full-scale fluid channel [m^2]
$x^h_{ch,SD}$	Hydraulic entrance length of a scaled-down fluid channel [m]
$x^h_{ch,FS}$	Hydraulic entrance length of a full-scale fluid channel [m]
Q_{SD}	Heat generated in the copper conductors of a scaled-down coil [W]
Q_{FS}	Heat generated in the copper conductors of a full-scale coil [W]
V_{SD}	Volume of the copper conductors in a scaled-down coil [m^3]
V_{FS}	Volume of the copper conductors in a full-scale coil [m^3]
q^m_{SD}	Mass flow rate in a scaled-down fluid channel [kg·s^{-1}]
q^m_{FS}	Mass flow rate in a full-scale fluid channel [kg·s^{-1}]
ρ	Fluid density [kg.m^{-3}]
C_P	Fluid specific heat capacity [J·kg^{-1}·°C^{-1}]
ΔT_{SD}	Fluid temperature difference in a scaled-down coil [°C]
ΔT_{FS}	Fluid temperature difference in a full-scale coil [°C]
$Pr_{ch,SD}$	Prandtl number in a scaled-down fluid channel [-]
x^t_{SD}	Thermal entrance length of a scaled-down fluid channel [m]

x^t_{FS}	Thermal entrance length of a full-scale fluid channel [m]
Q_{oil}	Volumetric oil flow rate measured by the flow Metre [m^3·h^{-1}]
P_1	Oil relative pressure measured in the bottom manifold [bar]
T_1	Oil temperature measured in the bottom manifold [°C]
$TC_1 - TC_{30}$	Acrylic temperatures measured [°C]
P_2	Oil relative pressure measured at the top manifold [bar]
T_2	Oil temperature measured in the top manifold [°C]
T_3	Oil temperature measured in the top pipe of the radiator [°C]
T_4	Oil temperature measured in the bottom pipe of the radiator [°C]
T_5	Ambient temperature [°C]
I_{supply}	Electrical current measured in the DC power supply unit [A]
I_{coil}	Electrical current measured by the multimeter in the coil [A]
V_{supply}	Voltage measured in the DC power supply unit [V]
V_{coil}	Voltage measured by the multimeter in the coil [V]
F_{pump}	Operating frequency of the gear pump [Hz]
F_{fan}	Operating frequency of the fan installed below the radiators [Hz]
P_{total}	Total power injected by DC power supply unit [W]
P_{coil}	Power dissipated in the coil [W]
R_{coil}	Ohmic resistance of the coil [Ω]
R_{total}	Total ohmic resistance of the circuit (includes cables and terminals) [Ω]
P_{cables}	Power dissipated in the cables [W]
$T_{avg,coil}$	Average temperature of the coil measured [°C]
T_{ref}	Reference temperature [°C]
$R_{coil,ref}$	Ohmic resistance of the coil measured at reference temperature [Ω]
P_x	Generic systematic uncertainty [units of the associated quantity]
$\bar{x}$	Generic arithmetic mean [units of the associated quantity]
B_x	Generic random uncertainty [units of the associated quantity]
U_x	Total combined uncertainty [units of the associated quantity]
$\overline{P_{coil}}$	Power density of the coil [kW·m^{-3}]
$\overline{u_G}$	Average oil velocity evaluated over the G section of the washer [cm·s^{-1}]
$P_{CFD,total,inlet}$	Total pressure at the inlet of the CFD scale model domain [Pa]
$P_{CFD,total,outlet}$	Total pressure at the outlet of the CFD scale model domain [Pa]
Δp^*_{CFD}	Total pressure difference in the CFD scale model domain [Pa]
Δp^*	Additional head loss. Estimated theoretically. [Pa]
$f_{b/t}$	Friction coefficient for the flow in the tubes of the manifolds. [-]
Δp_{CFD}	Total pressure difference (including the additional head loss) [Pa]
Δp_{EXP}	Total pressure difference measured in the experimental set-up [Pa]
$T_{avg,coil,CFD}$	Average temperature of the coil obtained in CFD [°C]
U	Global heat transfer coefficient of the coil [W·m^{-2}·°C^{-1}]
$A_{designed}$	Effective heat transfer area of the coil as designed [m^2]
$T_{avg,coil}$	Average temperature of the oil inside the coil [°C]

$T^*_{avg,coil}$	Average temperature of the coil obtained in CFD after area correction [°C]
$A_{manufactured}$	Effective heat transfer area of the coil as manufactured [m^2]
$TC_{CFD,x}$	Acrylic temperatures calculated in CFD for a generic position x [°C]
V_{ch}	Volume of the fluid channels [m^3]
H_{ch}	Height of the fluid channels [m]
L_{ch}	Characteristic length of the fluid channels [m]
$A_{w,ch}$	Wetted area of the fluid channels [m^2]
$A_{f,ch}$	Flow area of the fluid channels [m^2]
$d_{h,ch}$	Hydraulic diameter of the fluid channels [m]
P_n	Pressure in the fluid node n [Pa]
$q_{n-1:n}$	Mass flow rate in the fluid branch between the nodes $n-1$ and n [kg·s^{-1}]
$\Delta p_{n-1:n}$	Pressure drop in the fluid branch between the nodes $n-1$ and n [Pa]
$R^h_{n-1:n}$	Hydraulic resistance between the nodes $n-1$ and n [Pa·kg^{-1}·s]
$f(Re)$	Friction coefficient (function of the Reynolds number) [-]
$C^h_{n-1:n}$	Hydraulic conductance between the nodes $n-1$ and n [Pa^{-1}·kg·s^{-1}]
u_{ch}	Average fluid velocity in the fluid channels [cm·s^{-1}]
$Q^g_{i,j}$	Heat generated in the turn segment i,j [W]
$Q_{i,j,-X}$	Heat transferred/received from/to turn segment i,j along $-X$ [W]
$Q_{i,j,+X}$	Heat transferred/received from/to turn segment i,j along $+X$ [W]
$Q_{i,j,+Y}$	Heat transferred/received from/to turn segment i,j along $+Y$ [W]
$Q_{i,j,-Y}$	Heat transferred/received from/to turn segment i,j along $-Y$ [W]
$Q_{i,j,+Z}$	Heat transferred/received from/to turn segment i,j along $+Z$ [W]
$Q_{i,j,-Z}$	Heat transferred/received from/to turn segment i,j along $-Z$ [W]
nt	Total number of turns [-]
$ns(i)$	Number of turn segments associated with turn i [-]
$\Delta T_{i,j,+X}$	Temperature difference between turn segment i,j and the neighbouring turn segment along $+X$ [°C]
$R^t_{i,j,+X}$	Thermal resistance between turn segment i,j and the neighbouring turn segment along $+X$ [W^{-1}·°C]
$C^t_{i,j,+X}$	Thermal conductance between turn segment i,j and the neighbouring turn segment along $+X$ [W·°C^{-1}]
$R^t_{i,j,-Z}$	Equivalent thermal resistance between turn segment i,j and the adjacent fluid channel along $-Z$ [W^{-1}·°C]
$R^{fluid}_{i,j,-Z}$	Thermal resistance between the surface of the turn segment i,j and the adjacent fluid channel along $-Z$ [W^{-1}·°C]
$U^{fluid}_{i,j,-Z}$	Heat transfer coefficient between the surface of the turn segment i,j and the adjacent fluid channel along $-Z$ [W·m^{-2}·°C^{-1}]
$A^{fluid}_{i,j,-Z}$	Heat transfer area between the surface of the turn segment i,j and the adjacent fluid channel along $-Z$ [m^2]
$R^t_{i,j,-Y}$	Equivalent thermal resistance between turn segment i,j and the neighbouring turn segment along $-Y$ [W^{-1}·°C]

nb	Number of neighbouring turn segments along $-X$ [-]
nb^*	Number of neighbouring turn segments along $+X$ [-]
$T_{i,j}^c$	Temperatures in the turn segments [°C]
T_k^{fn}	Temperatures in the fluid nodes [°C]
nfn	Total number of fluid nodes [-]
$nconv$	Total number of fluid branches connected to a fluid node [-]
T_j^{fb}	Temperatures in the fluid branches [°C]
nfb	Total number of fluid branches [-]
X_F	Mass flow fraction in the fluid channels [%]
$q_{channel,i}$	Mass flow rate in the fluid channel i [kg·s^{-1}]
$q_{total,inlet}$	Mass flow rate at the inlet of the CFD model used for calibration [kg·s^{-1}]
RF_{ch}	Recirculation factor in the fluid channels [%]
$f_{T,CFD}$	Friction coefficient obtained from CFD for the transverse fluid channels [-]
$f_{R,CFD}$	Friction coefficient obtained from CFD for the radial fluid channels [-]
f_{Plates}	Analytical friction coefficient for infinite parallel plates [-]
$f_{4.24,Shah}$	Analytical friction coefficient for the ratio of the transverse fluid channels [-]
$f_{7.52,Shah}$	Analytical friction coefficient for the ratio of the radial fluid channels [-]
Nu_T	Nusselt number obtained from CFD for the transverse fluid channels [-]
Nu_R	Nusselt number obtained from CFD for the radial fluid channels [-]
Nu_{Plates}	Analytical Nusselt number for infinite parallel plates [-]
$Nu_{4.24,Shah}$	Analytical Nusselt number for the ratio of the transverse fluid channels [-]
$Nu_{7.52,Shah}$	Analytical Nusselt number for the ratio of the radial fluid channels [-]
$T_{ch}^{fn,FluSHELL}$	Temperature in the fluid channels from FluSHELL [°C]
$T_{ch}^{mwa,CFD}$	Average mass-weighted temperature in the fluid channels from CFD [°C]
$q_{channel}^{FluSHELL}$	Mass flow rate in the fluid channels from FluSHELL [kg·s^{-1}]
$q_{channel}^{CFD}$	Mass flow rate in the fluid channels from CFD [kg·s^{-1}]

Greek Letters

ϕ	Magnetic flux [Wb]
θ_A	Ambient temperature [°C]
θ_b	Bottom oil temperature [°C]
θ_o	Top oil temperature [°C]

θ_w Average winding temperature [°C]

θ_h Hot-spot temperature [°C]

φ_1 Global scaled-down factor of the experimental set-up [-]

φ_2 Spacers scaled-down factor in the experimental set-up [-]

σ Standard deviation [units of the associated quantity]

θ Dimensionless oil temperature [-]

Chapter 1
Introduction

Shell-type power transformers consist of a less known transformer technology comparing with the mainstream core-type power transformers.

The research work reported in this thesis concerns the development of a novel thermal model that is expected to provide means for better design (and exploration) decisions. This global objective has been achieved by applying a well-known numerical approach—based in thermal-hydraulic network analogies—and by validating its predictions against more detailed numerical approaches as well as with measurements. This novel thermal model is so far focused in a unitary system representative of the windings (the coil/washer system).

This is the introductory chapter and it has been subdivided in 5 sections:

1. Sect. 1.1 describes the importance of power transformers in the electrical grids worldwide;
2. Sect. 1.2 details in a top-down style each component of the shell-type transformers. This detailed decomposition of the transformer in its basic components intends on one hand to focus the main challenge addressed by this work but also sets the main nomenclature/definitions used throughout the thesis;
3. Sect. 1.3 describes the technological and economical motivations driving the need to develop such a detailed thermal-hydraulic algorithm. Along this section the relevance of this algorithm is articulated with other pertinent related areas of knowledge, namely the need to better understand and control the main ageing mechanisms influencing the end-of-life of transformers;
4. Sect. 1.4 describes the expected goals for this research work and give an adequate perspective of what has been accomplished and what is still part of future work;
5. Sect. 1.5 explains how the thesis has been organized as well as its main contents.

© Springer International Publishing AG 2018

H. Campelo, *FluSHELL – A Tool for Thermal Modelling and Simulation of Windings for Large Shell-Type Power Transformers*, Springer Theses,

https://doi.org/10.1007/978-3-319-72703-5_1

1.1   Background

According to IEC 60076-1 standard definition a transformer is *a static piece of apparatus (no moving components) with two or more windings which, by electromagnetic induction, transforms a system of alternating voltage and current into another system of alternating voltage and current usually of different values and at the same frequency for the purpose of transmitting electrical power* (IEC 2011a). In other words, each transformer receives energy at a certain voltage level in its primary circuit and delivers energy at a different voltage level from its secondary circuit. For simplification purposes, the primary and secondary circuits, can be understood to correspond to a Primary and a Secondary Winding. With some exceptions (e.g. autotransformers), in most cases the windings are not physically connected. Despite this, an alternating magnetic flux, ϕ, is guided through a high-permeability steel that creates an inductive link between the windings—this high permeability structure is the Magnetic Core. Whenever this magnetic field changes, proportional electromotive forces are observed in the terminals of both windings. This fundamental operating principle is one of the basic laws of electromagnetism and derives from Faraday's observation in 1831 (Wikipedia 2016a, b).

The induced voltage in the secondary winding, might be higher or lower than the voltage in the primary winding, depending on whether the transformer is designed for stepping up or stepping down the voltage level.

This flexible capability of transformers to modify voltage levels, together with the first public demonstrations of Alternating Current (AC) generators have influenced the course of the War of Currents in 1892 and since then the AC electrical grids became a worldwide standard up to nowadays (Uppenborn 1889).

The first transformer with a toroidal closed core dates back to 1885 and is attributed to a well-known group of three Hungarian engineers from Ganz factory in Budapest (Guarnieri 2013). Ever since, according to late L. F. Blume, transformers having been acting as major factor of economic development worldwide enabling the interconnection of different components throughout electrical grids. *Without this unique ability of the transformers to adapt the voltage to the individual requirements of the different parts of a system, and to maintain substantially constant voltage regardless of the magnitude of the load, the enormous development and progress in the transmission and distribution of electric energy, during the past 60 years, would not have been possible* (Blume et al. 1951).

The topology of an electrical grid varies worldwide and is continuously evolving. Across the electrical grids the transformers exist ubiquitously in different locations and with different expected functions. Transformers may exist:

- Near the heaviest generation sites (e.g. Nuclear Power Plants, Coal Plants and Hydro-Electric Plants). This region is usually denominated Transmission Grid. In this region of a grid, the transformers are usually connected to generators that produce energy at low voltage levels, between 10 and 40 kV (Del Vecchio et al. 2001). Then the transformers are used to step up the voltage level before electricity is fed into the network with the purpose of being transmitted over long distances at high voltages (typically higher than 220 kV).

- In the interconnections of the grid, where the grid progressively approximates the distribution level (typically below 110 kV) or where the grid needs to accommodate additional medium sized generation sites. At this level the transformers might also be useful to deliver energy to high-voltage consumers such as heavy industrial plants.
- Near the major consuming sites such as city or rural networks. This region is usually denominated Distribution Grid. More recently, a diverse range of renewable energies are being integrated at this voltage level which is modifying the classical hierarchized topologies with generation sites distant from the consumer sites. This is one of the key aspects behind the concept of Smarter Grids and this will shape the future expectations about the performance of transformers (Comission 2010).

Each electrical grid includes and combines several transformers with different sizes and types. According to IEC 60076-7 (IEC 2005), the transformers are classified according to their rated equivalent energy in MVA:

- a 3-phase transformer with a rated power up to 2.5 MVA is a distribution transformer;
- a 3-phase transformer with a rated power up to 100 MVA is a medium power transformer;
- a 3-phase transformer exceeding 100 MVA is a large power transformer.

In terms of construction, the transformers are classified according to the relative position between the windings and the transformer magnetic core:

- if the windings are wounded around the transformer core, the transformers are defined as core-type;
- if the transformer core encloses the windings, the transformers are defined as shell-type.

Nowadays, most of the manufacturers worldwide produce core-type transformers. In addition, several customers demand core-type in their technical specifications which are mainly supported on historical reasons and the inherent body of knowledge acquired through years of experience operating such equipment.

However, it is a matter fact that some of the major manufacturers in the transformer industry have also been manufacturing a significant number of shell-type transformers around the world for more than 100 years (some of them include both types of transformer technologies in their portfolio). The major players include companies such as Westinghouse and McGraw-Edison (Cooper) in USA, Jeumont-Schneider in France, ACEC in Belgium, ABB in Spain, IEM in Mexico, Hyosung in South Korea, Mitsubishi (MELCO) in Japan and more recently EFACEC in Portugal. Along this period some of these companies have been restructured or have disappeared, namely Westinghouse in USA from where a significant body of knowledge about this technology derives. For these reasons the global market share of this technology has been gradually lowering. Notwithstanding, there is a significant number of units being manufactured nowadays and the technology still has a high reputation due to its long-term resilience.

There no available public figures, but a total number of more than 25 000 shell-type power transformers are estimated to have been delivered worldwide so far. Among this total number:

- more than 15 000 transformers are estimated to have been delivered to the USA which corresponds to the biggest power market in the world;
- more than 3000 transformers are estimated to have been delivered to domestic customers in Japan;
- more than 7000 transformers are estimated to have been delivered in Europe namely for Belgium, France, Spain and Portugal. In Europe, it is noteworthy that all the nuclear fleet in Belgium, half of the nuclear fleet in France, more than 85% of the 400 kV network transformers in Spain and 100% of the 220 kV network transformers in Portugal are shell-type.

Along this period of 100 years some of these units might have reached its end-of-life or failed. Thus, if 80% of this total population is considered active, a total number of more than 20 000 shell-type transformers might be currently in service over electrical grids worldwide.

It is noteworthy that most shell-type transformers are located preferentially in Transmission Grids having on average a rated equivalent energy higher than

200 MVA/unit.

1.2   Shell-Type Transformers

A commercial power transformer, either core-type or shell-type, comprises a closed cooling loop as shown in Fig. 1.1.

Figure 1.1 shows an external perspective of a 700 MVA shell-type power transformer manufactured in 2012 at the EFACEC plant located in Savannah, USA. The active internal components are immersed in a large fluid volume enclosed in a steel Tank with an upper smaller Expansion Reservoir that ensures that the system is under constant pressure as it is designed to accommodate the fluid volume changes resulting from thermal expansions. The most common type of fluid used is a mineral naphthenic oil, which acts both as an internal cooling fluid and electrical insulator. The transformer shown weights approximately 450 tons, the steel tank is 10 metres high, the oil volume is approximately 30 cubic metres and oil circulation is imposed using 6 centrifugal pumps in parallel located at the Bottom Admission Circuit. According to IEC 60076-2 standard guidelines (IEC 2011b) this is

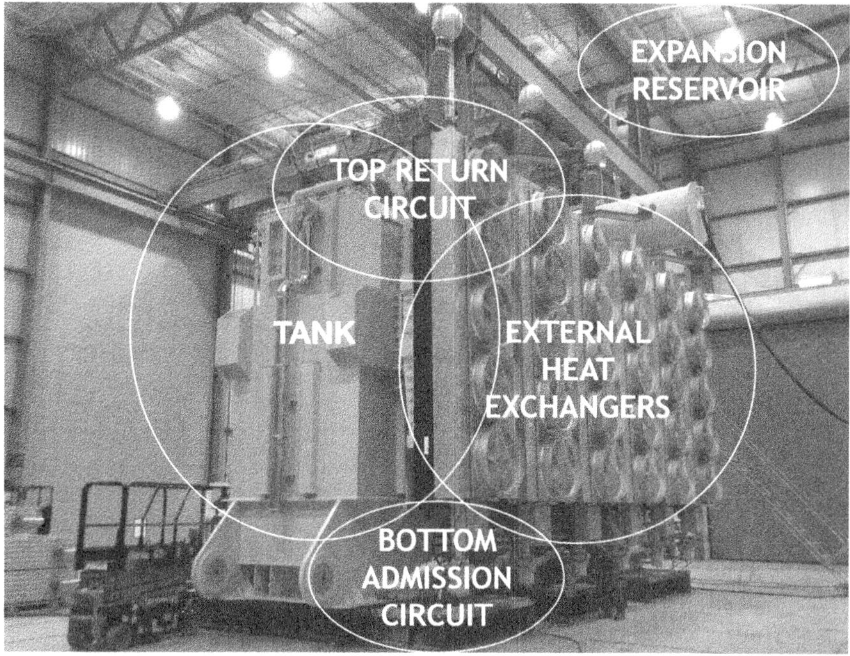

Fig. 1.1 Identification of the main components of a transformer cooling loop. External view of a commercial shell-type transformer. EFACEC Courtesy (Campelo 2015a)

classified as an Oil Distributed (OD) cooled power transformer—ODAF or ODAN, which would depend on the operating conditions.

The operation of such equipment is highly efficient from thermodynamics point of view. A large power transformer may exhibit efficiencies higher than 99.5%, although the remaining 0.5% can correspond to significant amount of energy heat being continuously generated and transferred to the internal cooling fluid. Under steady-state conditions, that same amount of generated is removed from the system using External Heat Exchangers. The Tank and these Heat Exchangers are connected through a Top Return Circuit where hotter oil coming from the tank arrives. After exchanging heat with ambient air, the colder oil is again re-admitted to the transformer and the whole cooling loop is repeated.

There are two major types of heat exchangers used in power transformers and both are shown in Fig. 1.2.

The whole transformer behaves thermally as a first-order system with a time constant in the range of few hours, namely due to the inertia of the large oil volume where its main components are immersed. For this reason, the IEC 60076-2 standard guidelines recommend temperature rise tests with durations of more than 5 h until a temperature variation below 1 $^\circ$C.h^{-1} is observed (IEC 2011b).

Thermodynamically it entails a closed cooling loop operating at constant pressure wherein the internal cooling fluid is incompressible. The flow regime inside

Fig. 1.2 Identification of the two major types of external heat exchangers. EFACEC Courtesy (Campelo 2015a)

transformer is mainly laminar and the equipment is designed to operate below acceptable temperature limits (as listed in IEC 60076-2).

Figure 1.3 depicts a large shell-type power transformer being commissioned in Seville, Spain. In this specific case of the transformer photographed the external cooling equipment is a group of vertical plate radiators with fans installed below them.

As above referred, the steel tank acts as an enclosure where all the active components of the transformer are kept immersed in naphthenic mineral oil. The main components of a shell-type transformer are shown schematically in the cut view of Fig. 1.4.

According to Fig. 1.4 the main components that can be found in shell-type transformers are the windings, the laminated magnetic core, the T-beams, the magnetic shunts and the external cooling equipment.

It is noticeable in Fig. 1.4 that, contrarily to core-type transformers, the distance between the internal tank walls and the laminated magnetic core is reduced. In this region, there are only few fluid channels to guarantee a physical separation between the steel of the magnetic core and the tanks walls as well as to guarantee an adequate evacuation of the heat generated in this region during operation. For this

Fig. 1.3 Shell-type transformer being commissioned in Seville, Spain. EFACEC Courtesy (Campelo 2015a)

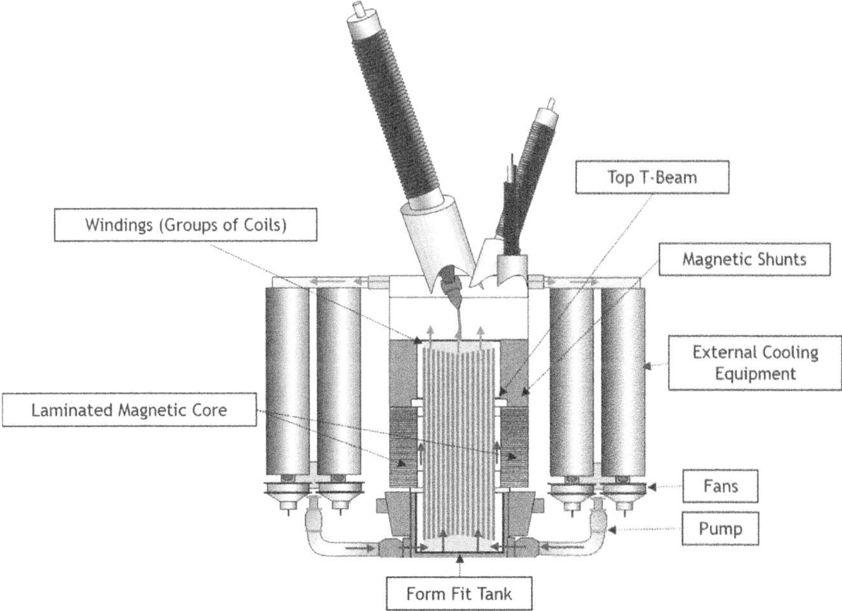

Fig. 1.4 Cut view of the main components of a shell-type transformer

reason, the tank is referred to be form fit. This characteristic implies less degrees of freedom for the cold oil re-entering the bottom tank.

Moreover, the weight of the magnetic core is supported in a steel structure called T-Beam, which is in turn supported in the re-entrant internal surfaces of the bottom tank—Fig. 1.4. The T-Beam together with the magnetic core creates a bottom pool of oil through which the oil is preferentially directed to the windings. In a core-type transformer this T-Beam structure would be like the tie plates typically located along each vertical limb of the magnetic core. Although, in a core-type equivalent transformer this bottom pool of oil would be larger and with significantly different hydraulic characteristics.

According to recent Computational Fluid Dynamics (CFD) results reported and compiled by the Working Group (WG) A2.38 of the International Council on Large Electric Systems (CIGRE), the oil expands suddenly after entering the bottom tank and thus a homogeneous pressure at the entrance of each coil seem to be an adequate assumption. This greatly simplifies the complex modelling of the thermal performance of each winding (Cigre 2016). In core-type transformers, this is not necessarily the case. According to a survey from the CIGRE WG 12.09 which has been conducted among utilities spread worldwide, 19 core-type transformers out of a total of 33 did not exhibit any particular system to guide the oil in the bottom tank to the windings (Cigre 1995). For this and other reasons the IEC60076-2 standardizes six different cooling modes for fluids with thermal class lower than 300 °C (IEC 2011b).

1. Oil Directed Air Forced with acronym ODAF;
2. Oil Directed Air Natural with acronym ODAN;
3. Oil Forced Air Forced with acronym OFAF;
4. Oil Forced Air Natural with acronym OFAN;
5. Oil Natural Air Forced with acronym ONAF;
6. Oil Natural Air Natural with acronym ONAN.

The two first cooling modes, ODAF and ODAN, refer to designs where the oil is pumped and directed (or guided) to the windings. The difference between these two cooling modes, concerns the ambient air and whether it is forced to circulate through the external heat exchangers by using fans or not (AF or AN, respectively). The next two cooling regimes OFAF and OFAN, refer to designs where the oil is pumped but no structures exist in the bottom tank to preferentially direct the oil to the windings. Finally, ONAF and ONAN cooling modes refer to designs where the oil is not pumped, instead the flow is buoyancy driven.

For the specific case of shell-type transformers, although not standardized, the list of cooling modes might be simpler. Due to the technological characteristics above discussed, whenever pumps exist the shell-type transformers are intrinsically ODAF or ODAN.

Each one of the components depicted in Fig. 1.4 is detailed below with emphasis on the windings which are the focus of this work.

1.2.1 Windings

Each winding or, more precisely, each part of a winding (Low-Voltage—LV or High-Voltage—HV), is composed of coils. The coils are represented in Fig. 1.4 by the thin solid rectangles disposed vertically.

Each winding is composed by alternating groups of coils. As each group of coils is not arranged consecutively the whole arrangement is referred as being interleaved. This arrangement is depicted in Fig. 1.5.

As the equivalent power of a shell-type transformer increases, the shape of each coil remains identical. Instead of modifying the geometry of the coils, the ampere-turns are reduced by introducing additional coils. This maintains the magnitude of the electromechanical forces independent of the size of the transformer and creates parallel thermal-hydraulic circuits with similar hydraulic resistances. For instance, in a core-type transformer the hydraulic resistance of each winding might be quite different (e.g. a tertiary or a regulation layer-type winding without guides compared with a typical guided disc-type winding) and additional design decisions must be assumed to compensate that. An interesting example can be found in (Cigre 2016) where perforated bottom pressboard structures are reported to be used as oil flow distributors in the bottom tank.

The form fit tank combined with almost uniform hydraulic resistances, between the windings and between each coil, comprise the main reason why the research efforts are herein focused in single copper coils. At this moment, this is believed to comprise the most relevant and representative unitary domain of the windings, while it can be confidently decoupled of the upstream and downstream conditions.

In large power transformers, it is common, that more than 80% of the heat is generated inside the coils. The coil is expected to be one the highest stressed component inside the transformer, according to a recent reliability survey conducted on over 964 transformers, the windings were identified as one of the major cause of failures in substation transformers with voltages higher than 100 kV (Cigre 2015). Each coil is formed by a variable number of copper conductors through

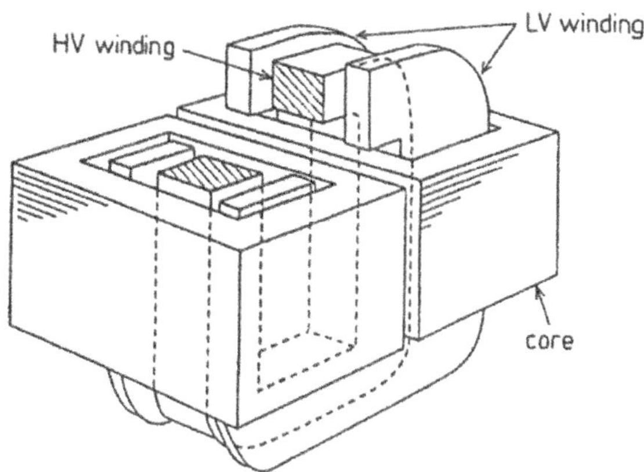

Fig. 1.5 Interleaved winding arrangement in a shell-type transformer. Image from (Campelo 2015b)

Fig. 1.6 Photo of two shell-type coils during manufacturing stage. Schematic representation of a single bundle. Images from (Campelo 2015b)

which the main alternating electrical current is circulated. In the case depicted in Fig. 1.6 five individual copper conductors with rectangular cross sections are bundled together to form a turn. The electrical current circulates in parallel amongst the five conductors of each bundle. Then each bundle of five conductors is wounded around in several turns to form the pancake shaped coil photographed. Due to a superimposition of inductive and resistive effects, energy is dissipated under the form heat inside each copper conductor of each coil (Del Vecchio et al. 2001).

The capability of modelling the electromagnetic induced losses and its spatial distribution is beyond the scope of the current research work and the heat has been always considered as a boundary condition imposed uniformly as a source in each single copper conductor (Cigre 2016). This a common procedure to decouple effects. In addition, there is a manifest difficulty in isolating the heat generated due to resistive and inductive effects, so the experiments reported in this thesis have been conducted under DC conditions, which means the heat is uniformly distributed over the coil and heat is generated exclusively due to resistive effects.

For simplification purposes, the geometry of the coil used in this work comprises a turn (or bundle as above referred) with a single copper conductor wounded around 48 times which corresponding to 48 turns.

For a complementary internal perspective, Fig. 1.7a includes a longitudinal cut view of a three phase shell-type power transformer in order to emphasise the pressboard pieces used to create the fluid channels adjacent to the surfaces of each coil.

Each coil is sandwiched between two washers made of high-density pressboard with trapezoidal shaped spacers distributed and glued over it—Fig. 1.7b. These structures are commonly called spacers and are used to open fluid channels through which the internal cooling medium circulates. The volume and cross section area opened between the pressboard washer and the copper coil surface defines the fluid channels through which mineral oil flows while the spacers' height define fluid channel height (typically between 4 and 6 mm)—Fig. 1.8.

These trapezoidal shaped spacers also guide the internal cooling fluid that wets the heated coil surfaces, hence removing energy from them. Moreover the location and number of these spacers must be balanced in terms of mechanical withstanding capability and heat transfer area covered (Campelo 2015b).

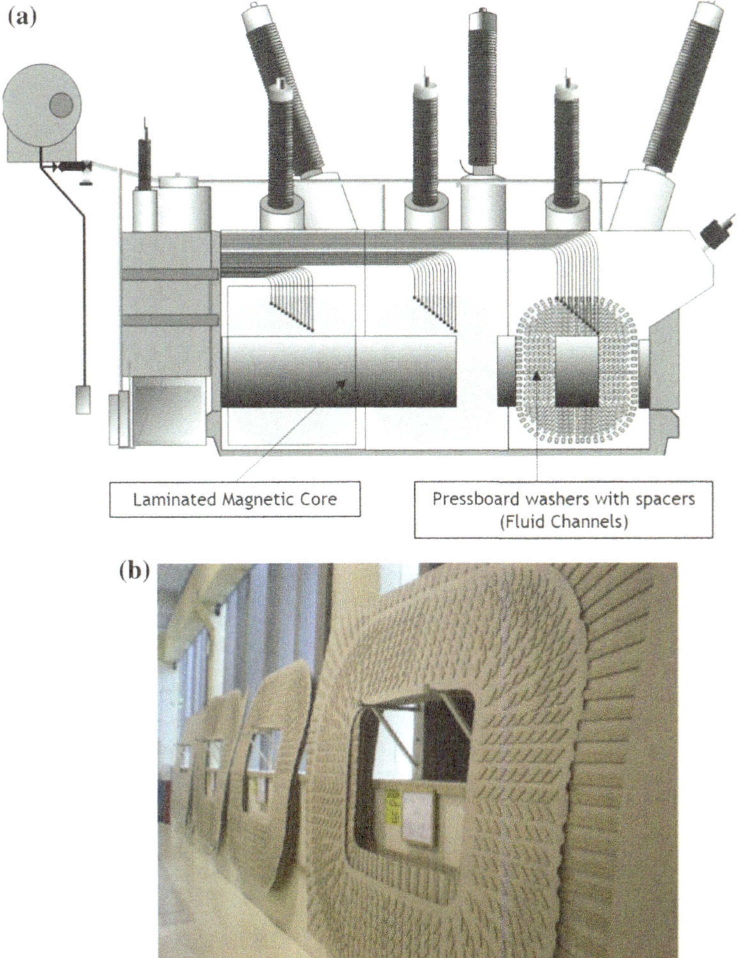

Fig. 1.7 **a** longitudinal cut view of a shell-type transformer and **b** pressboard washers with spacers before being assembled. Images from (Campelo 2015b)

Then each coil is stacked-up as in Fig. 1.9 and the spacer's location must be coincident from bottom to top to transmit forces homogeneously guaranteeing effective mechanical stability of the whole phase.

A crucial component of each coil are the insulation frames which are folded around the innermost and outermost turns for electrical reasons. These are protective elements which also confer some mechanical stability to final stack of coils. The insulation frames are also made of high-density pressboard and are moulded in order produce shapes as those shown in Fig. 1.10a and they might be assembled as shown in Fig. 1.10b. From a thermal-hydraulic point of view this is one of the most distinctive characteristic of this transformer technology and they are of upmost

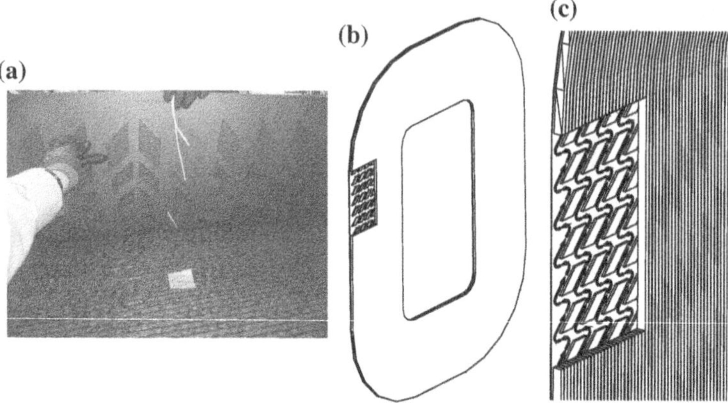

Fig. 1.8 Shell-type coil and adjacent pressboard washer with spacers glued over it: **a** photograph **b** schematic representation and **c** zoom emphasizing the fluid channels with oil circulating. Images from (Campelo et al. 2015b)

Fig. 1.9 Stack of coils. Complete assembly of one single phase. Images from (Campelo 2015b)

relevance in any thermal modelling approach as they represent an effective reduction of the coil heat transfer area and hence represent one of the locations where higher temperatures are expected.

Fig. 1.10 Insulation frames to fold around the innermost and outermost turns of each coil: **a** before assembling and **b** after assembling in a commercial coil. Images from (Campelo 2015b)

The copper coil, the washer, the spacers and the insulation frames form together the coil/washer system. This is the system where this work is focused.

As these structures are folded around the outermost and around the innermost turns of each coil, they create additional restrictions to heat transfer to oil around these turns.

When designing a coil of shell-type transformer some details must be considered to open fluid channels that increase the wetted area around these regions. The insulation frames can be folded around the two surfaces of a coil or around one of them only (U-Shaped insulation frames or L shaped insulation frames as reported in (Cigre 2016)). Considering the early stage of maturity, these dented-like structures shown in Fig. 1.10 have been primarily approximated in this work by a more basic shape corresponding a linear pressboard strip that completely covers the wetted area available in the innermost and outermost turns. At this stage, this has served to decompose the influence of these structures to better understand the combined impact of such structures.

The insulation frames are believed to be the highest thermally stressed region in each coil. The parametric influence of the shape of these structures on the thermal performance of shell-type coils is certainly a topic of future relevance and interest. Currently it has been important to develop a thermal model that is sensible to the main heat transfer mechanisms acting in these special regions of each coil.

1.2.2 Laminated Magnetic Core

The magnetic core results from stacking laminations of electrical steel around the windings. A schematic representation is shown in Fig. 1.11 where the yellow coloured homogeneous blocks intend to represent several laminations of electrical steel stacked together (with a thickness between 0.23–0.30 mm).

As most of the metals, this steel is a polycrystalline solid composed by several magnetic domains. The superior magnetic permeability of this material ($\approx$1500 times higher than air) guarantees that the main magnetic flux is conducted through the magnetic core. Although, under the presence of alternated magnetic fields, the boundaries of the magnetic domains move which generate dissipation of energy under the form of heat (Del Vecchio et al. 2001).

As a result, fluid channels need to be also opened in this region of the transformer to evacuate this heat while maintaining the steel surface temperatures below critical temperatures of 140 °C, as recommended in IEC 60076-7 (IEC 2005).

The capabilities of modelling the losses generated in the core and the corresponding temperatures are beyond the current scope of this work.

1.2.3 T-Beams and Magnetic Shunts

Even though the main magnetic flux is *guided* through the magnetic core, there are magnetic fluxes linking other structural components inside a shell-type transformer. These magnetic fluxes are commonly called leakage fluxes and they are also responsible for generating additional heat in other components (Penabad-Duran et al. 2014; Sitar et al. 2015).

Figure 1.12a depicts a group of perpendicular magnetic shunts positioned both in the internal tank walls and below the lower T-Beam. Figure 1.12b depicts an additional group of parallel magnetic shunts located in the other internal tank walls.

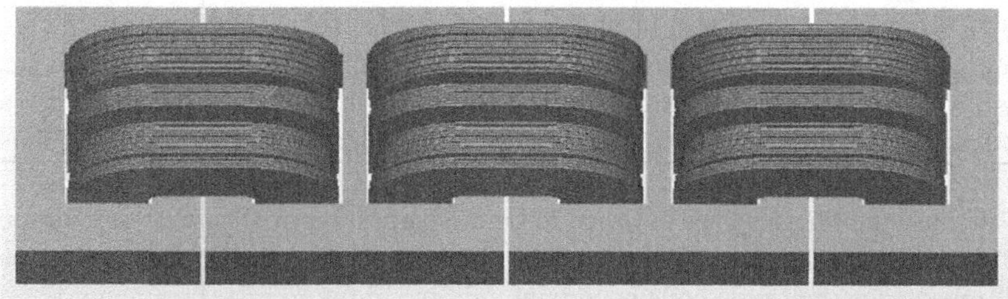

Fig. 1.11 Top view of the magnetic core embracing the windings of a 3-phase shell-type transformer. Image from (Campelo 2015b)

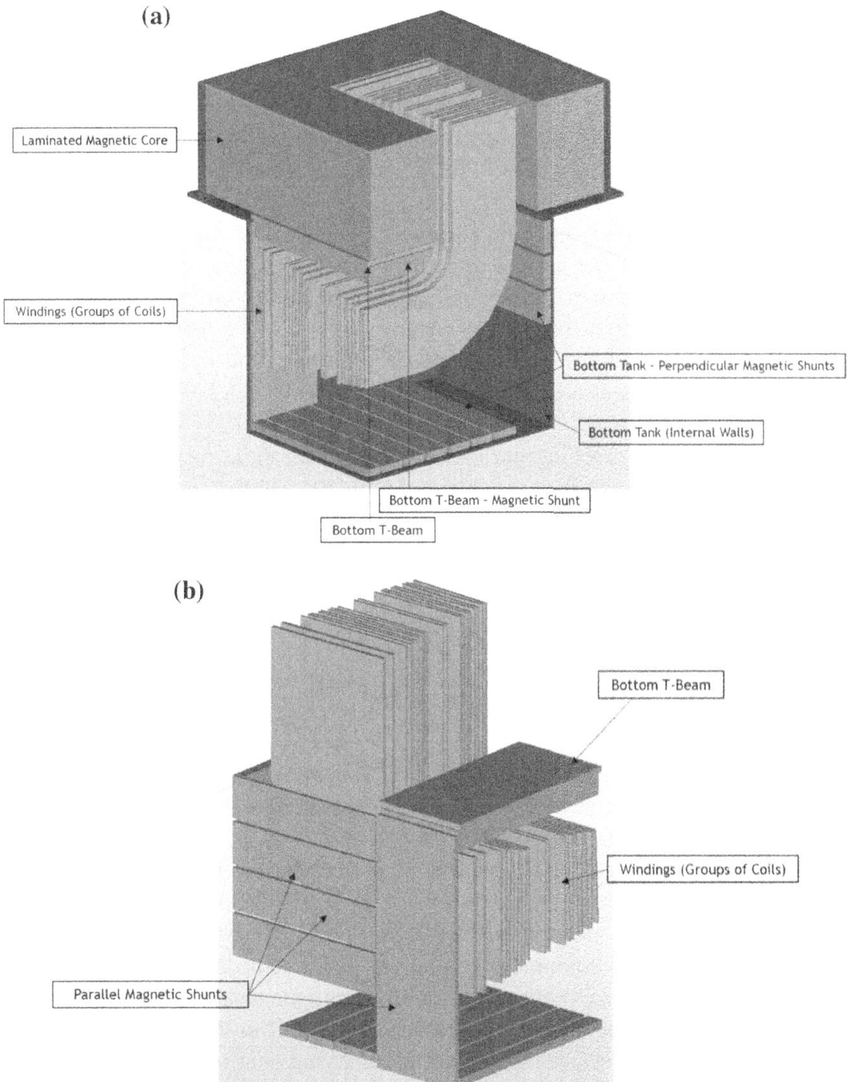

Fig. 1.12 Images of typical magnetic shunts located inside a shell-type transformer: **a** perpendicular magnetic shunts and **b** parallel magnetic shunts. Images from (Campelo 2015b)

The magnetic shunts are classified as perpendicular and parallel according to their relative position compared with the incident magnetic field. The thickness of these structures is designed to limit induction in these surfaces below saturation (which occurs at $\sim 2T$).

The T-Beams are metallic structures, used exclusively in shell-type transformers that help to maintain the windings in a vertical position while simultaneously withstanding the weight of the core and the windings. These structures are also protected using magnetic shunts (Cigre 2016).

As a consequence of the incident alternating magnetic field, there is also additional heat generation in these magnetic protective structures, which demands additional fluid channels to efficiently remove this energy and maintain the surface temperatures below critical values (140 °C as recommended in IEC60076-7 (IEC 2005)).

The oil circulation in the fluid channels opened near the lower and upper T-beams is assumed to not influence the heat transfer conditions of the innermost turn of each coil. Although there is still not much public information about this technology and this might be a topic of future interest. The most comprehensive descriptions about this transformer technology can be found in (Lopez-Fernandez et al. 2012) and in a dedicated chapter of the CIGRE WG A2.38 Brochure (Cigre 2016).

The capability of modelling the losses generated in these magnetic shunts and the corresponding temperatures are assumed to be beyond the current scope of this work. The description of these components intends to give a broader idea of whole thermal-hydraulic related phenomena involved in the design of a large power transformer.

1.2.4 External Cooling Equipment

The external cooling equipment shown in Fig. 1.3 and in Fig. 1.4 is attached to the tank and exists to guarantee that the energy generated under the form of heat (in the active components mentioned) is removed to the ambient air (AN cooling regimes as defined in IEC60076-2 (IEC 2011b)). The ambient air is the most common external cooling medium, although in some special applications water can be used (WF cooling regimes as defined in IEC60076-2 (IEC 2011b)).

The external cooling equipment is attached to the tank through top pipes and bottom pipes. The top pipes collect the hot oil exiting the tank (red coloured arrows in Fig. 1.4) and the bottom pipes re-introduce colder oil in the bottom part of the tank (blue coloured arrows in Fig. 1.4) after removing heat to the ambient air.

Depending on the design, certain transformers might have axial or centrifugal pumps located either in the bottom pipes or in the top pipes of the external cooling equipment. These pumps are used to impose a constant pressure that forces the circulation of the internal cooling medium, hence increasing the heat evacuation capability of the internal cooling medium (OD cooling regimes as defined in IEC 60076-2 (IEC 2011b)).

Identically, and again depending on the design, the external cooling equipment might include fans to force the circulation of the external cooling medium. This also increases the heat evacuation capability of these equipment (AF regimes as defined in IEC 60076-2 (IEC 2011b)).

Due to this interconnected combined performance of the external cooling equipment together with the active components immersed in the tank, the whole system is understood to comprise a closed cooling loop. Whenever a transformer is energized heat starts being generated all over the metallic components of the transformer (copper conductors in the coil, magnetic core, magnetic shunts, etc.) and the temperatures globally start to increase. After a certain amount of time these temperatures tend to stabilize whenever the external cooling equipment is able to evacuate to the ambient air the same amount of heat being generated internally. This corresponds to the instant where the whole system is understood to have achieved thermodynamic equilibrium and it is herein defined as steady-state.

In this work, the need for modelling the external cooling equipment has been avoided by fixing the oil temperature entering each coil as a boundary condition. The thermal modelling approaches used address exclusively a single coil behaviour under steady-state equilibrium. The next section describes in detail the main motivations driving the need to develop increasingly accurate and more detailed tools to predict the temperature distribution inside each copper coil.

1.3 Motivation

Temperature is one of the most relevant parameters driving the ageing and limiting the loading capability of an electric transformer. For transformers with solid insulation designated as class 105 °C and immersed in mineral oil the average and maximum temperatures during operation are limited according to IEC60076-2 (IEC, 2011b): the top oil temperature must not exceed 80 °C, the average winding temperature of an OD transformer must not exceed 90 °C and the maximum temperatures in the windings must not exceed 98 °C. These limits refer to steady-state conditions under continuous rated power and under a yearly average ambient temperature of 20 °C.

Each transformer is designed to meet these temperature limits and the details of its design are closely exchanged with the customers along comprehensive design review meetings even before initiating the purchase of any component (Cigre 2013).

At the end of manufacturing, and apart from specific agreements, every new transformer undergoes a strict testing sequence, in a High Voltage Laboratory. Usually this testing is conducted in the manufacturing plant where the transformer has been manufactured. The testing sequence might include electrical, dielectrical, mechanical, thermal and chemical tests. The methodologies underpinning each specific test can be found in IEC 60076-2 (IEC 2011b) or in IEEE C57 12.90 (Board 2006). The test sequence aims to assess the overall quality of the equipment and commonly includes thermal tests. During these tests—often defined in the standards as temperature rise tests—the measured temperatures are compared against the mentioned guaranteed temperature limits.

In some cases, these temperature rise tests are complemented by on-site tests when the transformer is commissioned (Tanguy et al. 2013).The manufacturer usually guarantees the normal operation of the equipment for a period not lower than 5 years. This provides an insight about how the risks in this industry are balanced among each stakeholder and how the stakeholders share common interests over the whole life-cycle of the equipment.

Acceptable temperature limits, for normal load and for overload conditions, are standardized respectively in IEC 60076-2 (IEC 2011b) or in IEC 60076-7 (IEC 2005), and are based on the influence of temperature in the ageing mechanisms of the solid insulation as well in the ageing mechanisms of the cooling fluid. Under these temperatures both solid insulation and cooling fluid are expected to degrade normally while contributing to an acceptable lifetime of the asset under operation in the electrical grid. A transformer is expected to be in service between 30 to 50 years (Oliva et al. 2010).

Moreover, as early reported by Montsinger, the lifetime of a transformer is determined by the magnitude of its highest temperatures. Which means, that the ideal temperature distribution should be as uniform and low as possible. Montsinger also quantified that, for every 6 °C increase in the highest temperature of a winding, the lifetime of a transformer is halved (Montsinger 1930).

According to a survey conducted by the CIGRE WG 12.09 (Cigre 1995), from 1995, 17 out of 27 utilities define its overload policies according to the overload capability of the transformers. Again, a transformer with a better thermal performance under its rating power will enable an optimized overload capability (beyond its nameplate rating power) and therefore will optimized management policies (as for example, delayed investments or extended lifetime of the assets) (Picher et al. 2010).

Consequently, the capability predicting and controlling temperatures with accuracy has been driving the research efforts of both transformer manufacturers and electrical utilities worldwide (Campelo et al. 2016; Picher et al. 2010; Tanguy et al. 2004). From the perspective of a manufacturer, these capabilities are expected to empower the engineering teams with better design tools that can generate improved design decisions while from the perspective of a utility these capabilities are expected to induce better management decisions and lower failure rates of the equipment (Cigre 2015).

It is noteworthy that, under some circumstances, the decision whether to remove or not a transformer from the electrical grid—End-Of-Life decision—is based on ageing evidences and hence indirectly based on temperature. These ageing evidences might arise from laboratorial analysis of solid insulation samples taken directly from the transformer or intelligently guessed using ageing models (Martins et al. 2011).

The ageing of a transformer is a chemical reaction through which the solid insulation as well as the fluid degrades and increases the probability of failure of the transformer. The typical solid insulation used in transformers is made from pure Kraft pulp due to two main characteristics: its excellent oil impregnation capability and its good geometric stability in oil. After refinement processes, the typical composition of the unbleached softwood Kraft pulp is 78–80% cellulose, 10–20% hemicellulose and 2–6% lignin (Krause et al. 2014).

In turn, cellulose is a linear polymer and its degradation is quantified by measuring the average size of its macromolecules using viscometric methods. This indicator is called the viscometric degree of polymerization, DP_v. The DP_v of the solid insulation in a new unaged transformer is ≈ 1200 and the transformer achieves its end-of-life when the same insulation reaches a final DP_v of 200. At this degree of polymerization, the cellulose is expected to maintain 50% of its initial tensile strength which decreases significantly the mechanical stability of the whole transformer.

Under operation the cellulosic components of a transformer are continuously exposed to chain scissions promoted by three main chemical reactions:

- hydrolysis which is governed by the water content and catalysed by carboxylic acids;
- oxidation which is governed by the oxygen content and catalysed by hydroxyl radicals;
- And pyrolysis which is governed exclusively by temperature. At normal operating temperatures below 140 °C this mechanism is expected to be less relevant.

Simultaneously, the naphthenic mineral oil typically used is also continuously exposed to an oxidation reaction that results in the formation of additional carboxylic acids, which in turn accelerates the main hydrolysis reaction of the cellulose.

All these chemical reactions obey an Arrhenius type law, and the activation energy associated with each reaction is dependent on temperature. So the capability of predicting temperatures accurately if of upmost importance in order to better understand and control this complex ageing processes (Cigre 2007).

Besides this, the temperature also influences the electrostatic charging tendency observed in the fluid/solid interfaces. This tendency increases with temperature and is also highly dependent on the oil flow Reynolds number, being indeed one of the main reasons for operating the transformer under laminar flow regimes. This phenomenon is commonly called static (or streaming) electrification and, in order to prevent it, the local oil velocities in the fluid channels of the windings are generally limited to maximum values of 1 m.s^{-1} (Moser et al. 1992). Having referred this, it is relevant to mention, that the oil velocity distribution inside a transformer is not uniform and that localized high velocity regions usually bottlenecks the design of the whole transformer by avoiding the use of higher average oil flow rates. Ultimately, an optimal cooling efficiency is understood to depend on a more uniform oil velocity distribution.

In this sense, the capability of predicting accurately the oil velocity distribution inside the fluid channels of a winding can significantly impact the design of the transformer. This is one of the additional added values of using detailed algorithms such as thermal-hydraulic network models and CFD. These algorithms not only predict the temperature distribution but also predict how the oil flow is actually distributed between each fluid channel (Campelo et al. 2012).

Despite this, it is also interesting to acknowledge that besides the mentioned economic reasons driving the need for advanced prediction methodologies, significant drivers may also be attributed to major developments in the sensing systems as well as to increasingly fast and affordable computational capabilities (Campelo et al. 2013).

Since early days, the thermal performance of transformers has been assessed experimentally with the four types of measurements as depicted in the conventional thermal diagram of Fig. 1.13: ambient temperature,θ_A, bottom oil temperature, θ_b, top oil temperature, θ_o, and the average winding temperature, θ_w. The first three values are measured directly with thermal devices and the latter is obtained indirectly using the winding resistance extrapolated to the instant of shutdown. Through a manipulation of these measurements, an average winding gradient, g, is calculated and afterwards multiplied by an empirical Hot-Spot Factor, H, to obtain the Hot-Spot Temperature, θ_h.

The empirical Hot-Spot Factor is used to account for the non-linear behavior of a transformer winding. This non-linear behaviour had been often attributed to the asymmetric distribution of electromagnetic losses, which are higher in the top discs where the incident magnetic field is expected to induce increased losses. The IEC 60076-7 (IEC 2005) refers typical Hot-Spot Factors of 1.3 for medium and large power transformers. Although the same IEC standard mentions that several studies reveal that this factor can range from 1.0 to 2.1. The IEC 60076-2 Annex B (IEC

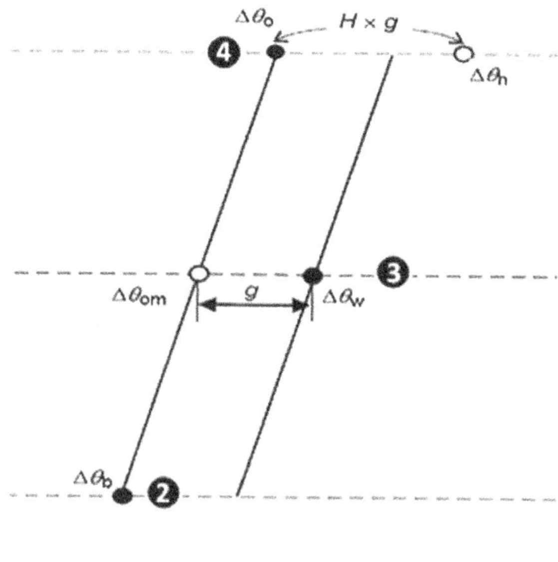

Fig. 1.13 Simplified thermal diagram of a transformer winding. Diagram adapted from IEC 60076-2 Annex B (IEC 2011b)

2011b), revised recently in 2011, recommends the separation of this Hot-Spot in two separate factors: the factor Q associated with the asymmetric distribution of electromagnetic losses and the factor S associated with the flow distribution. Hence the Hot-Spot Factor results from the multiplication of these two factors, H = QS. Anyhow, if the thermal model calculates directly the oil flow distribution in each channel and it is also able to account for a non-uniform loss distribution along the winding, these simplifications assume less relevance.

This historical background moved the efforts towards the prediction of the referred average winding gradients, g, which have been early understood as being a sum of two temperature drops:

1. the temperature drop experienced by the heat when flowing through the insulation of the conductors;
2. and the temperature drop experience by the heat when flowing through the surface of the insulated conductors and the oil;

The temperature drop through the insulation is a consequence of the thermal conduction through the insulation layers while the temperature drop through the surface is a consequence of a coupled conduction-convection in the oil side.

Consequently, these temperature drops have been lumped into mathematical expressions that have been built based on extensive experimental data extracted from commercial transformers. And according to (Cigre 2016) these expressions still comprise the most common method used to design the windings of shell-type transformers.

These thermal modelling approaches have been enlightening, namely during the 1930s and up to 1980s, due to the limited capabilities of predicting the position and magnitude of the highest temperature occurring inside the complex geometry of a transformer. At that stage, most of the engineering systems were studied analytically, such as heat exchangers for example (Shah and London 1971).

On the other hand, the embedded temperature sensors available at that period did not comply satisfactorily with the practical requirements of not weakening dielectrically the winding to be measured (Norris 1928). Which means that at this stage the measurement equipment was considered intrusive.

This resulted in empirically relevant thermal models, due to its high correlation with experimental data collected over years of experience, but physically less meaningful having many details lumped together. Such details can become particularly important whenever the hottest temperature position varies under different loading conditions and particularly when increasing efficiency is needed both from technological or economical perspectives (Picher et al. 2010; Tanguy et al. 2004).

Nevertheless, during the 1980s, the numerical techniques applied to complex engineering systems evolved rapidly pacing together with faster and affordable computers. During this decade some of the most relevant numerical methods to solve fluid flow and heat transfer problems have been further integrated in robust computational packages known generally as Computational Fluid Dynamics (CFD) (Patankar 1980). This promoted a new engineering paradigm based in virtual experiments rather than conventional costly actual experiments. This paradigm is still dominant nowadays and is subsequently driving the efforts of many researchers worldwide towards the development of increasingly detailed capabilities of prediction.

Simultaneously, the most recent sensor technologies developed so far have been proven to withstand high voltage stresses without weakening the measuring target (Nordman and Takala 2010). Emphasis for the fiber optic based temperature sensors that enabled a direct measurement of the hottest temperature occurring inside transformer windings, without the need to resort on empirical Hot-Spot Factors.

This evolving and intricate context has motivated previous applications of CFD and thermal hydraulic network modelling techniques to diverse design aspects of core-type transformers. In 2008, CFD was first used to model in detail the thermal-hydraulic behaviour of a group of windings from a commercial core-type transformer (Campelo et al. 2009). In 2012 a first thermal-hydraulic network model has been developed for the same type of transformers. This model has introduced the novelty of using calibrated expressions for friction and heat transfer coefficients that have been previously extracted from CFD (Campelo et al. 2012), the same principle being now applied to shell-type transformers. Between 2012 and 2014, the performance of this model was assessed through comparison with measurements collected during standard temperature rise tests conducted in commercial transformers. However, the measurements on commercial transformers pose serious challenges to rigorous thermofluid measurements. Besides the absence of precise mass flow rate measurements in each individual winding, the scale of the equipment and its characteristics can induce measurement variations that are not directly attributable to a single isolated parameter (Campelo et al. 2014a).

Thus, when comparing predictions with direct temperature measurements in commercial transformers some scatter can be expected due to a build-up of systematic and random uncertainties. The current challenge is then to be able to characterize the scatter and identify whether part of it is systematic, for example whether it derives from certain limitations of these techniques to model the reality, or whether it occurs for specific geometric configurations or for specific operating conditions. Since the mastering of these thermal modelling techniques is of common interest for both manufacturers and utilities, a long-term R and D collaboration framework has been initiated in 2014 between EFACEC and *Institute du Recherhe d'Hydro-Québec* (IREQ). Along this collaboration, the identical numerical approaches used by both institutions are being benchmarked and compared against tightly controlled experiments conducted in a scale model that represents a typical thermal-hydraulic circuit of a core-type transformer winding. The results so far indicate an adequate correlation between these modelling approaches and experiments (Campelo et al. 2016; Torriano et al. 2016).

In parallel, since 2011 an identical strategy has been put through also to shell-type transformers, with the particularity that the use of a scale model has been considered since earlier stages. At this moment, this thesis reports the development of a novel thermal-hydraulic network model—the FluSHELL tool—that is intended to improve the existing calculation methodologies for shell-type transformers while addressing some of the current and future challenges above described.

The next section describes the specific objectives of the work.

1.4   Objectives

The primary objective of this research was to develop a thermal-hydraulic network tool to aid transformer designers in the process of designing a shell-type transformer winding in the following ways:

- To provide means for the designer to compute the steady-state temperature distribution along a coil, instead of computing its average thermal gradient to oil.
- To provide means for the designer to compute the steady-state oil flow distribution inside the fluid channels wetting the surfaces of a coil, instead of computing the average fluid velocity for the whole coil.
- To provide means for the designer to compute the steady-state distribution of temperatures along a coil, based in fixed boundary conditions of either mass flow rate or total pressure.
- To provide means for the designer to evaluate changes due to variations in specific design parameters namely: the shape of the insulation frames, the position of the spacers, the geometric properties of the coil, the heat generation rate, the cooling fluid flow rate and the cooling fluid inlet temperature.
- To reduce significantly the time involved when comparing with an identical thermal-hydraulic simulation conducted using a commercial CFD code.

Besides this, the new tool also aims to aid transformer research engineers enriching the thermal design-cycle with more capabilities namely:

- To provide means for future proprietary integrations: with non-existent knowledge about the remaining components of the closed cooling loop (e.g. pumps, external cooling equipment, etc.), with other existing systems of information (where the needed inputs might be available under different forms) and with broader system-level tools or equipment (e.g. integration with other multidisciplinary algorithms or integration in monitoring systems).

As a complement to the objectives mentioned, a conceptual validation of the developed tool has been assumed critical, having the following set of specific objectives:

- To design and to manufacture a scale model representative of the closed cooling loop of oil immersed coil of a commercial shell-type power transformer.
- To conduct tightly controlled experiments in order to compare measured temperatures against predicted temperatures.

From an industrial perspective, this new tool is expected to exhibit improved sensitivity and accuracy comparing with existing methodologies, enabling more efficient design decisions and also more efficient root cause analysis (under contexts of failure for example). The comparison against measurements collected during tests in full-scale commercial shell-type power transformers (either on the HV Lab or on-site) comprises a further stage of maturity and is beyond the scope of the work.

As a broader pedagogical objective, this research work also aims to help disseminating the knowledge about a less known transformer technology and, whenever pertinent, some of results must be put in perspective against the mainstream core-type transformer technology.

1.5 Thesis Outline

Apart from this introductory Chapter, the remaining contents of this thesis are organized in 5 additional Chapters:

- Chapter 2—Scale Model, where a new experimental setup is described;
- Chapter 3—CFD Scale Model, where a detailed CFD Model has been developed to represent that setup and has been compared against experimental results;
- Chapter 4—the FluSHELL Tool, where a novel proprietary thermal tool is presented;
- Chapter 5—FluSHELL Validation where that new tool is validated against both experiments and CFD;
- Chapter 6—Conclusions and Future Work where the accomplishment of each objective is assessed and future activities motivated by this research work are envisaged

References

Blume, L. F., Boyajian, A., Camilli, G., Lennox, T. C., Minneci, S., & Montsinger, V.M. (1951). *Transformer engineering: A treatise on the theory, operation, and application of transformers*.

Board, I. -S. S. (2006). *C57.12.90: IEEE standard test code for liquid-immersed distribution, power, and regulating transformers*.

Campelo, H. M., Lopez-Fernandez, X. M., Picher, P., & Torriano, F. (2013). Advanced thermal modelling techniques in power transformers. Review and case studies. In: *Advanced Research Workshop on Transformers* (pp. 1–17). Baiona.

Campelo, H. M. R. (2015a). Cooling of large power transformers using CFD: Vision, strategy and case studies. In: *ANSYS Convergence Conference*. Porto.

Campelo, H. M. R. (2015b). Large shell-type power transformers: New perspectives through advanced thermal modelling techniques. In: *Weidmann Transformers Seminar*. Zurich.

Campelo, H. M. R., Baltazar, J. P. B., Oliveira, R. T., Fonte, C. M., Dias, M. M., & Lopes, J. C. B. (2015b). Extracting relevant transport properties using 3D CFD simulations of shell-type electrical transformers. In: *International Symposium on Advances in Computational Heat Transfer*. New Jersey, USA: Begell House.

Campelo, H. M. R., Braña, L. F., & Lopez-Fernandez, X. M. (2014a). Thermal hydraulic network modelling performance in real core type transformers. In: *2014 International Conference on Eletrical Machines (ICEM)* (pp. 2275–2281). Berlin, Germany: IEEE. doi:http://dx.doi.org/10.1109/ICELMACH.2014.6960502.

Campelo, H. M. R., Fonte, C. M., Lopes, R. C. L., Dias, M. M., & Lopes, J. C. B. (2012). Network modelling applied to CORE power transformers and validation with CFD simulations. In: *International Colloquium Transformer Research and Asset Management* (pp. 1–16). Dubrovnik.

Campelo, H. M. R., Fonte, C. M., Sousa, R. G., Lopes, J. C. B., Lopes, R. C., Ramos, J., Couto, D., & Dias, M. M. (2009). Detailed CFD Analysis of ODAF Power Transformer. In: *International colloquium transformer research and asset management* (pp. 1–10). Cavtat, Croatia.

Campelo, H. M. R., Quintela, M. A., Torriano, F., Labbé, P., & Picher, P. (2016). Numerical thermofluid analysis of a power transformer disc-type winding. In: *IEEE Electrical Insulation Conference*. Montreal, Canada.

Cigre. (2016). WG A2.38 brochure, transformers thermal modelling. Draft version 5.

Cigre. (2015). WG A2.37: Transformer Reliability Survey.

Cigre. (2013). WG A2.36: Guidelines for conducting design reviews of power transformers.

Cigre. (2007). WG Brochure 323: Ageing of cellulose in mineral-oil insulated transformers.

Cigre. (1995). WG 12.09 Brochure: Thermal aspects of transformers.

Comission, E. (2010). European electricity grid initiative roadmap and implementation plan.

Del Vecchio, R. M., Poulin, B., Fegahli, P. T., Shah, D. M., & Ahuja, R. (2001). *Transformer design principles: With applications to core-form power transformers*. Gordon and Breah Science.

Guarnieri, M. (2013). Who invented the transformer? *IEEE Industrial Electronics Magazine, 7.*

IEC. (2011a). IEC 60076: Power transformers—Part 1: General.

IEC. (2011b). IEC60076: Power transformers—Part 2: Temperature rise for liquid-immersed transformers.

IEC. (2005). IEC 60076: Power transformers—Part 7: Loading guide for oil-immersed power transformers.

Krause, C., Dreier, L., Fehlmann, A., & Cross, J. (2014). The Degree of polymerization of cellulosic insulation: Review of measuring technologies and its significance on equipment. In: *IEEE Electrical Insulation Conference*. Philadelphia, USA.

Lopez-Fernandez, X. M., Ertan, H. B., & Turowski, J. (2012). *Transformers: Analysis, design, and measurement*. CRC Press.

Martins, M. A., Fialho, M., Martins, J., Soares, M., Cristina, M., Lopes, R. C., & Campelo, H. M. R. (2011). Power transformer end-of-life assessment—pracana case study. *IEEE Electrical Insulation Magazine, 27.*

Montsinger, V. M. (1930). Loading transformers by temperature. *Transaction of AIEE*, 776–790.

Moser, H. P., Krause, C., Praxl, G., Spandonis, G., & Stonitsch, R. (1992). Influence of transformerboard and nomexboard on the electrification of power transformers. In: 3rd *EPRI Workshop*. San Jose, USA.

Nordman, H., & Takala, O. (2010). Transformer loadability based on directly measured hot-spot temperature and loss and load current correction exponents. In: *CIGRE Session*. Paris, France.

Norris, E. T. (1928). The thermal rating of transformers. *Journal of the Institution of Electrical Engineers, 66.* http://dx.doi.org/10.1049/jiee-1.1928.0095.

Oliva, M., Prieto, A., Cuesto, M., Fernandez, A., Porrero, J., & Jalinat, A. (2010). Large generator setp tup transformers with low hot spot for EDF nuclear power plants. In: *CIGRE Session* (p. A2_303_2010). Paris, France.

Patankar, S. V. (1980). *Numerical fluid flow and heat transfer*. CRC Press.

Penabad-Duran, P., Lopez-Fernandez, X.M., & Turowski, J. (2014). 3D non-linear magneto-thermal. *Electrical Power Systems Research* doi:http://dx.doi.org/10.1016/j.epsr. 2014.11.010.

Picher, P., Torriano, F., Chaaban, M., Gravel, S., Rajotte, C., & Girard, B. (2010). Optimization of transformer overload using advanced thermal modelling. In: *Cigré Conference* (p. A2_305_2010). Paris.

Shah, R. K., & London, A. L. (1971). *Laminar flow forced convection heat transfer and flow friction in straight and curved ducts, a summary of analytical solutions*. California: Stanford University, Mechanical Enginnering.

Sitar, R., Janic, Z., & Stih, Z. (2015). Improvement of thermal performance of generator step-up transformers. Applied Thermal Engineering. doi: http://dx.doi.org/10.1016/j.applthermaleng. 2014.12.052.

Tanguy, A., Patelli, J. P., Devaux, F., Taisne, J. P., & Ngnegueu, T. (2004). Thermal performance of power transformers: Thermal calculation tools focused on new operating requirements.In: *Cigré Conference* (p. A2_105_2004).Paris.

Tanguy, A., Ryadi, M., Channet, J., & Hurlet, P. (2013). In service experience of hot spot behaviour of a GSU power transformer compared to temperature rise test results. In: *Cigre A2&D1 Joint Colloquium*. Kyoto, Japan.

Torriano, F., Campelo, H. M. R., Labbé, P., Quintela, M. A., & Picher, P. (2016). Experimental and numerical thermofluid study of a disc-type transformer winding scale model. In: *XXIInd International Conference on Eletrical Machines*. Lauzanne, Switzerland.

Uppenborn, F. (1889). History of the transformer.

Wikipedia. (2016a). Electrical transformer [WWW Document]. URL https://en.wikipedia.org/ wiki/Transformer.

Wikipedia. (2016b). Electrical grid [WWW Document]. URL http://en.wikipedia.org/wiki/ Electrical_grid.

Chapter 2
Scale Model

An experimental setup has been designed, manufactured and used to validate directly the Computational Fluid Dynamics (CFD) and the FluSHELL tool results. The experimental setup is a scaled down model incorporating the physical representation of the closed cooling loop occurring in a commercial full-scale shell-type coil. This model combines, in a controlled environment, the main thermal-hydraulic characteristics of interest for this specific transformer technology.

The use of scale models is of paramount importance as powers transformers are units possibly weighting hundreds of tons, where some components might have meters of length and, at the same time, where millimetric variations must be perceived, understood and controlled to ensure a proper performance. Before being commissioned, the quality of these units is extensively controlled in laboratory tests, which include among others tests to assess whether the temperatures are below or not the guaranteed limits. Although extremely useful for acceptance or not of the equipment, extensive and rigorous research based exclusively in full-scale tests might be insufficient (or at least costly), as the performance is a cumulative result of a diverse range of parameters.

The use of scale models represents a highly-valuable approach to better control and isolate variables of interest and therefore enabling increasingly robust decisions when comparing the experiments with numerical prediction capabilities.

The Sect. 2.1 describes the state-of-the-art concerning experimental validations of thermal modelling techniques for power transformers in general, either core-type or shell-type. Experimental validations have been widely used in transformer industry, but most of them have been focused in core-type transformers. Most experiments have been used to validate different prediction capabilities.

In Sect. 2.2 the experimental setup is described. Firstly, the scaling-down considerations used to ensure representativeness of the model are detailed. Afterwards the components of the setup are detailed emphasizing dimensions as well as the equipment used for acquisition and measurements.

In Sect. 2.3 the methods used to ensure reproducible, steady-state measurements are described and some results are presented.

© Springer International Publishing AG 2018

H. Campelo, *FluSHELL – A Tool for Thermal Modelling and Simulation of Windings for Large Shell-Type Power Transformers*, Springer Theses,

https://doi.org/10.1007/978-3-319-72703-5_2

Finally, in Sect. 2.4, global conclusions about this scale model approach are listed.

2.1 Introduction

As described in Chap. 1, a power transformer, either core-type or shell-type, comprises a closed cooling loop where the heat generated in the coils is removed to the ambient using external heat exchangers. Along Chap. 1 it has also been described that each transformer is commonly tested before being commissioned. And, in principle, this could be thought to generate enough experimental data to be used to validate all the capabilities of prediction.

Although, there are some practical limits to tests in power transformers:

1 The scale of the equipment poses serious challenges to a rigorous installation of sensors and in some cases, is it not possible to install adequate sensors. For example, it is not common to measure the oil flow rate in each coil or even winding;
2 The manufacturing process introduces additional uncertainties such as materials and other dimensional tolerances;
3 In some cases, the insufficient number of reproducible units to test (this limitation might be dependent on the scale of the manufacturer).

Thus, it is difficult to isolate the effect of each variable, and tests in power transformers are not sufficient to validate the physical assumptions underlying mathematical models such as CFD and FluSHELL. Therefore, significant research efforts have been conducted over the years to build more controlled experimental environments (Campelo et al. 2015).

One of the first known experiments has been conducted by Higgins at the National Physical Laboratory in 1928 (Higgins and Davis 1928). At the time a closed cooling loop was built to assess the most relevant physical properties influencing heat transfer around a coil. The coil was simplified by using a helicoidal platinum resistance and the focus was the cooling fluid. Results showed that dynamic viscosity was the most influent parameter in the thermal performance of the coil. In 1958 a former employee of C.A. Parsons, UK reported one of the most comprehensive experiments so far (Taylor et al. 1957). In this work, several arrangements of coils, common in core-type transformers, were built and local temperatures measured. Empirical formulations to predict the thermal performance of such type of coils were developed. In 1982 Szpiro built an experimental apparatus of a similar coil arrangement and measured the maldistribution of the oil flow, suggesting that the arrangement of the oil ducts inside the coil greatly influence its thermal performance (Szpiro et al. 1982). Contrarily to Taylor, these experiments were conducted using a Perspex model to visualize the oil flow and anemometers were used to measure exclusively the local oil flow distribution. More recently, in

2008, in a work supported by VA Tech (Siemens), Zhang reported the development of a similar scale model also made of Perspex where different cooling duct arrangements have been tested (Zhang et al. 2008). In this model the axisymmetric effect has been first introduced using tapered spacers. The oil temperatures measured in each and the disc temperatures have been used to extract Nusselt correlations to be further implemented in a thermal-hydraulic network model that was then validated. In 2009 Weinlader reported the development of a scale model identical to Szpiro with the added value of measuring temperatures and velocities simultaneously. These latter experiments from Weinlader were used to validate 3D CFD simulations (Weinlader and Tenbohlen 2009; Weinlader et al. 2012). All the experiments referred are said to have been conducted in scale models not due its scaled down dimensions, but since they comprise fewer number of disks than those existing in a full-scale coil.

The most recent scale model, which already includes the whole cooling loop, has been reported by Torriano et al. (2016). This scale model adds further value due to the representation of the whole cooling loop targeted for buoyancy driven, natural convection flows (without pumps forcing the circulation of oil). The results of this scale model were used to validate both CFD and network modelling approaches. It is noteworthy that these efforts are being driven by Hydro-Québec which is one of the main utilities in North America, reinforcing the idea that this knowledge is relevant either from a manufacturer or user perspective. In this case the knowledge built upon these experiments is expected to be used to assess whether some transformers can be safely overloaded and thus optimizing significant investments in new assets (Picher et al. 2010).

All the experiments mentioned above refer to core-type coil arrangements. In 2007, a similar experiment to Szpiro was conducted, but this time for a typical shell-type coil arrangement (Gomes et al. 2007), where Particle Image Velocimetry measurements were used to validate the isothermal flow distribution predicted using 3D CFD simulations. This last contribution, is the most important for the scope of this research work, and can be understood to precede the current experimental setup that is herein described.

In the current experimental setup, a coil has been added and hence the coil/washer system (with external heat exchangers included) can be validated under non-isothermal conditions with possibility of measuring directly average and local temperatures.

This experimental setup is comprehensively described in the next section.

2.2   Experimental Setup

The experimental setup designed and built is based upon the geometric dimensions of a commercial 700 MVA shell-type low-voltage coil and intended to be representative of the closed cooling loop, which comprises the admission and return

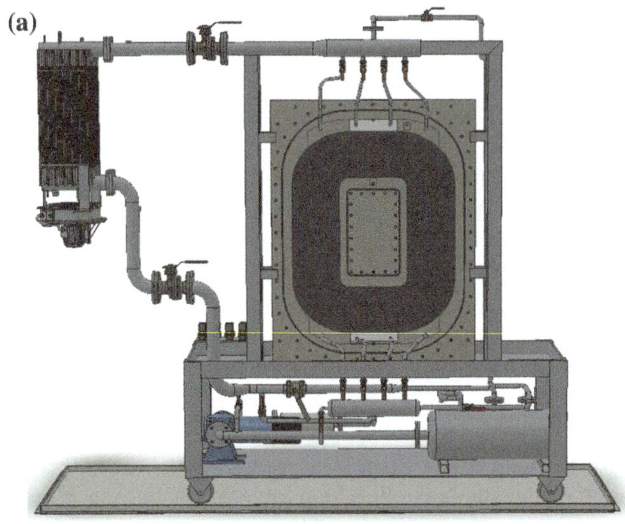

Fig. 2.1 Experimental setup: **a** schematic 3D drawing and **b** actual setup. EFACEC Courtesy

circuits as well as the heat exchanger. The setup focused on a single coil as shown in Fig. 2.1.

To maintain the setup within reasonable dimensions, and whenever feasible, the full-scale dimensions have been multiplied by a scaled down factor of $\varphi_1 = 1/3$. The exception to this rule concerns the height of the spacers which have been scaled down by a factor of $\varphi_2 = 1/2$ (due to the commercial unavailability of thinner acrylic plates). The main implications of using these scale-down factors are addressed along the next section.

2.2.1 Scaling-Down Considerations

Because of the scaled down geometry, certain criteria have been defined for both the fluid velocities in the fluid channels and the heat dissipated in the coil.

2.2.1.1 Fluid Velocities

To maintain similarity between the velocity fields, it has been defined that the same average velocity magnitude must be maintained between the full-scale fluid channels and the scaled down fluid channels, or

$$u_{ch,SD} = u_{ch,FS} \tag{2.1}$$

Where $u_{ch,SD}$ is the average velocity in a scaled down fluid channel and $u_{ch,FS}$ is the average oil velocity in a full-scale fluid channel. This in turn implies a relationship between the Reynolds number of the scaled down geometry and the Reynolds number of the full-scale geometry according to:

$$Re_{ch,SD} = \frac{d_{h,ch,SD}}{d_{h,ch,FS}} Re_{ch,FS} \approx \varphi_2 Re_{ch,FS} \tag{2.2}$$

Where $Re_{ch,SD}$ is the Reynolds number in a scaled down fluid channel, $d_{h,ch,SD}$ is the hydraulic diameter of a scaled down fluid channel, $d_{h,ch,FS}$ is the hydraulic diameter of a full-scale fluid channel, $Re_{ch,FS}$ is the Reynolds number in a full-scale fluid channel and φ_2 is the scale factor between the height of the spacers in the scaled down geometry and the full-scale geometry.

Hence, for this experimental setup, the relationship between scaled down and full-scale Reynolds number is

$$Re_{ch,SD} \approx \frac{1}{2} Re_{ch,FS} \tag{2.3}$$

And thus, the flow regime is laminar in both geometries.

Moreover, to maintain similar average fluid velocities, the relationship between the volumetric flow rates is given by

$$q^v_{ch,SD} = \frac{A_{f,ch,SD}}{A_{f,ch,FS}} q^v_{ch,FS} = \varphi_1 \varphi_2 q^v_{ch,FS} \tag{2.4}$$

Where $q^v_{ch,SD}$ and $q^v_{ch,FS}$ are the volumetric flow rates for a scaled down fluid channel and for a full-scale fluid channel, respectively, and, $A_{f,ch,SD}$ and $A_{f,ch,FS}$ are the respective flow areas.

Therefore, for this experimental setup, the relationship between scaled down and full-scale volumetric flow rate is

$$q^v_{ch,SD} = \frac{1}{6} q^v_{ch,FS} \tag{2.5}$$

This has no meaning besides the demand for a smaller pump in the current experimental setup.

In addition, the relationship between the hydraulic length of a scaled down fluid channel and the hydraulic length of a full-scale fluid channel is:

$$x^h_{ch,SD} = \frac{Re_{ch,SD}d_{h,ch,SD}}{Re_{ch,FS}d_{h,ch,FS}} x^h_{ch,FS} \tag{2.6}$$

Where $x^h_{ch,SD}$ and $x^h_{ch,FS}$ are the hydraulic entrance lengths of a scaled down fluid channel and a full-scale fluid channel, respectively.

Thus, for this experimental setup, this relationship is:

$$x^h_{ch,SD} = \varphi_2^2 x^h_{ch,FS} \Leftrightarrow x^h_{ch,SD} = \frac{1}{4} x^h_{ch,FS} \tag{2.7}$$

which means that in the scaled down fluid channels the required length for the flow to develop is 25% of the expected length for the same full-scale fluid channels. The hydraulic lengths of a representative scaled down geometry have been analysed in Chap. 4 where two types of fluid channels were described, the transverse fluid channels with an average hydraulic length of 4.5 mm and the radial fluid channels with an average hydraulic length of 3.2 mm. According to this relationship, the corresponding average hydraulic lengths in the full-scale are 18 and 12.8 mm, which represents less than 50% of the full-scale channel lengths. Therefore, both in the scaled down and in full-scale geometry, the flow is expected to be fully developed over most part of the channels.

2.2.1.2 Heated Dissipated in the Coil

To maintain similarity between the temperature fields, it has been defined that the same heat dissipation per unit of volume should be maintained between the full-scale coil and the scaled down coil

$$\frac{Q_{FS}}{V_{FS}} = \frac{Q_{SD}}{V_{SD}} \Leftrightarrow Q_{SD} = Q_{FS} \frac{V_{SD}}{V_{FS}} \tag{2.8}$$

Where Q_{FS} and Q_{SD} are the heat dissipated in a full-scale coil and in a scaled down coil, V_{FS} and V_{SD} are the volumes of the corresponding full-scale coil and scaled down coil. As all the dimensions of the full-scale coil have been multiplied by the scaled down factor φ_1, the relationship between the volumes of the scaled down and the full-scale coil is given by

$$\frac{V_{SD}}{V_{FS}} = \varphi_1^3 \Leftrightarrow \frac{V_{SD}}{V_{FS}} = \frac{1}{27} \tag{2.9}$$

thus, enabling the usage of a power supply equipment with a more convenient size:

$$Q_{SD} = \frac{1}{27} Q_{FS} \tag{2.10}$$

Besides the size of the power supply, some implications are expected in the fluid temperatures.

Since the system is practically adiabatic and operates at constant pressure, the fluid enthalpy balance between the inlet and outlet of the coil is given by

$$Q = q^m \rho C_P \Delta T \tag{2.11}$$

where q^m is the fluid mass flow rate, ρ is the fluid density, C_P is the fluid specific heat at constant pressure and ΔT is the fluid temperature difference between the outlet of the coil and the inlet of the coil.

Assuming this, the relationship between the fluid temperature difference in a scaled down coil and in a full-scale coil is determined by

$$\frac{\Delta T_{SD}}{\Delta T_{FS}} = \frac{Q_{SD}}{Q_{FS}} \frac{q^m_{FS}}{q^m_{SD}} \tag{2.12}$$

Furthermore, if the ratio between volumetric flow rates is similar to the ratio between mass flow rates, the final relationship between fluid temperature differences inside the coil is given by

$$\Delta T_{SD} = \frac{2}{9} \Delta T_{FS} \tag{2.13}$$

which means that, maintaining the same heat dissipation by unit of volume and the same fluid velocities, results in a fluid temperature increase in the coil that is approximately 22% of the fluid temperature increase in the corresponding full-scale coil.

Moreover, the thermal entrance length is described by the following equation:

$$x^t_{ch,SD} = x^h_{ch,SD} Pr_{ch,SD} \tag{2.14}$$

where $Pr_{ch,SD}$ is the fluid Prandtl number in a scaled down fluid channel. Consequently, assuming the Prandtl number is similar in a scaled down fluid channel and in a full-scale fluid channel, the relationship between the thermal lengths in a scaled down fluid channel and in a full-scale fluid channel is identical to the relationship between hydraulic lengths

$$x^t_{SD} = \frac{1}{4} x^t_{FS} \tag{2.15}$$

which means that in the scaled down fluid channels the required length to achieve a thermal developed profile is 25% of the expected length for the same full-scale fluid channels. The thermal lengths of a representative scaled down geometry have been also analysed in Chap. 4, where transverse fluid channels are reported to have an average thermal length of 446 mm while the radial fluid channels are reported to have an average thermal length of 323 mm. The average geometric length of a scaled down transverse channel 23.3 mm while in the radial channels is 13.3 mm, so these thermal lengths still largely exceed the geometric channel lengths. Therefore, both in the scaled down and in full-scale geometry, the flow is expected to be thermally developing.

2.2.2 Description of Experimental Setup

The experimental setup and its main components are depicted in Fig. 2.2.

The main overall dimensions of the experimental setup, in mm, are indicated in Fig. 2.2. Table 2.1 complements this information identifying each numbered component of the setup (Table 2.2).

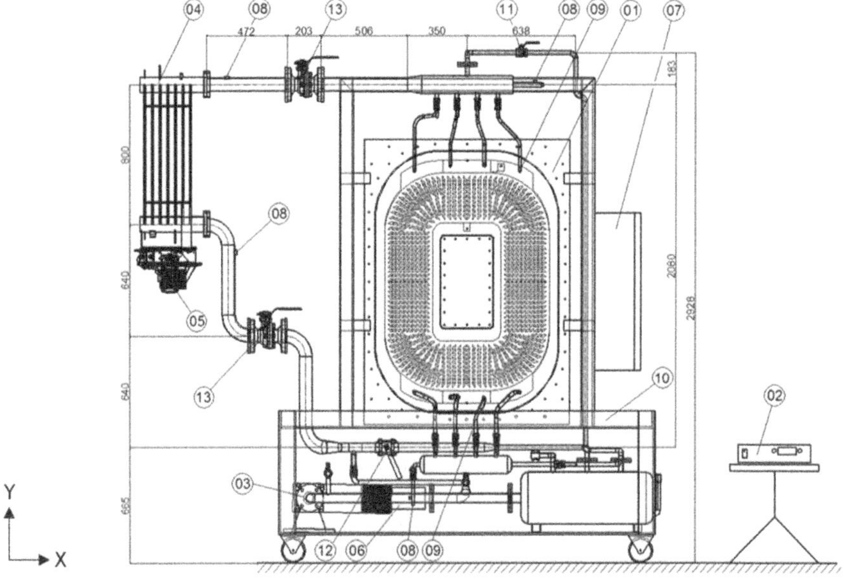

Fig. 2.2 Experimental setup (excluding the upper expansion reservoir and simplifying minor details). Dimensions in mm

Table 2.1 Identification and description of the main components of the experimental setup in Fig. 2.2

ID	Component
01	Coil/Washer system (copper coil sandwiched between two acrylic plates—detailed view in Fig. 2.3)
02	DC power supply and data acquisition (Power supply—Agilent® N8754 A–20 V/250 A–5000 W) (Data acquisition—HP xw4400 workstation)
03	Pump with frequency controller (Efaflu® Gear Pump—Motor 3F IE2) (Frequency controller with 2.2 kW/950 rpm)
04	Vertical plate radiator (Warm® S.L.—7 plates—800 mm x 520 mm)
05	Radiator fan (Efaflu® fan—180 W/400 V/Diameter 300 mm–1350 rpm)
06	Ultrasonic flowmeter (In-line model—Sonoelis® SE 404X–DN32)
07	Control cabinet (Siemens Simatic® panel with data recording and interface to control pump and fan)
08	Control sensors (with 4–20 mA current outputs): Pressure—2 X IFM® PA3024 (ceramic capacitive electronic sensor—relative pressure cell) Temperature—4 X IFM® TA3130 (platinum resistance thermometers type PT1000 class A)
09	Bottom manifold (cylindrical steel container with 1 inlet and 4 outlets regulated with individual globe valves). The top manifold is identical
10	Supporting steel structure with wheels
11–13	Other globe valves (details about the complete circuit in Fig. 2.4)

Table 2.2 Sub-components of the coil/washer system identified in Fig. 2.3

ID	Component
01.01	Acrylic plates (made of PMMA each with 20 mm along Z)
01.02	Acrylic spacers (made of PMMA with 2 mm along Z)
01.03	Inner and outer insulation frames (cover completely the 1st and 48th turns—except in the inlet and outlet regions)
01.04	Coil with 48 turns. Each turn includes a copper bar with 8.8 mm × 2.8 mm covered with Denisson 22 HCC with 0.588 mm thickness (the coil is 9.976 mm along Z). Each turn is separated by pressboard strips with 2 mm thickness
01.05	Fabor® Toric rings with 6 mm diameter (made of nitrile rubber)
01.06	Coil terminal (made of brass—where each lead of the coil is brazed)
01.07	Pressboard blocks surrounding the copper coil

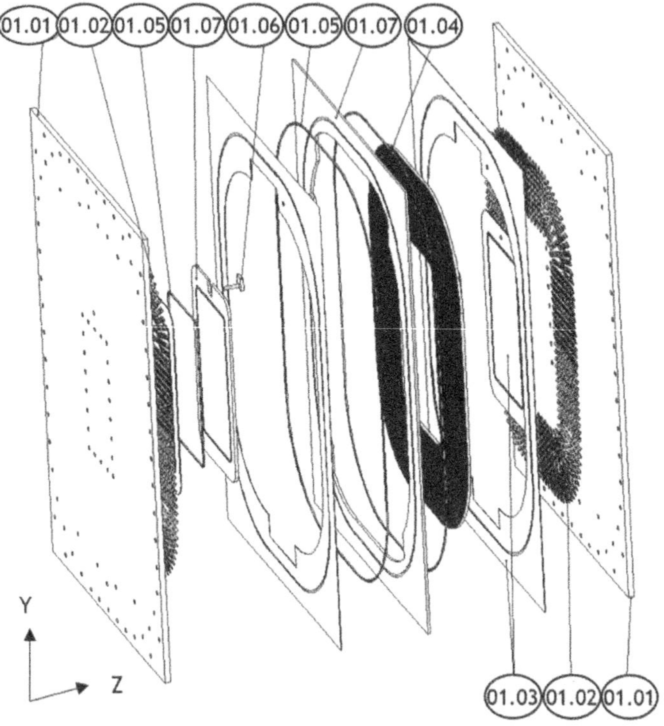

Fig. 2.3 Detailed view of the coil/washer system in the experimental setup (along the Z coordinate)

Figure 2.3 further details the sub-components of the coil/washer system in a detailed view along the Z coordinate.

Figure 2.3 shows that the coil is sandwiched between two acrylic plates (01.01). The acrylic plates are shown perforated to allow the introduction of tightening bolts. In commercial transformers, this system is expected to be compressed by the weight of other coils or components such as the magnetic circuit. In the experimental setup, this is achieved with an adequate tightening of the coil.

The spacers which create the fluid channels (01.02) have been bonded to these acrylic plates using Extrufix® solvent from Bostik®. The copper coil (01.04) has been enclosed using high density pressboard blocks around it (01.07) hence decreasing the heat losses to the ambient air while maintaining the coil under nominal dimensions. These blocks have the same dimension as the copper coil along Z (9.976 mm). Additional pressboard structures (01.03) are glued on the top of that blocks to cover the innermost and outermost turns of the copper coil. These additional structures represent the inner and outer insulation frames and have the same dimension of the spacers along the Zcoordinate—2 mm.

Toric rubber rings have been used in the pressboard blocks to prevent oil leakages (01.05).

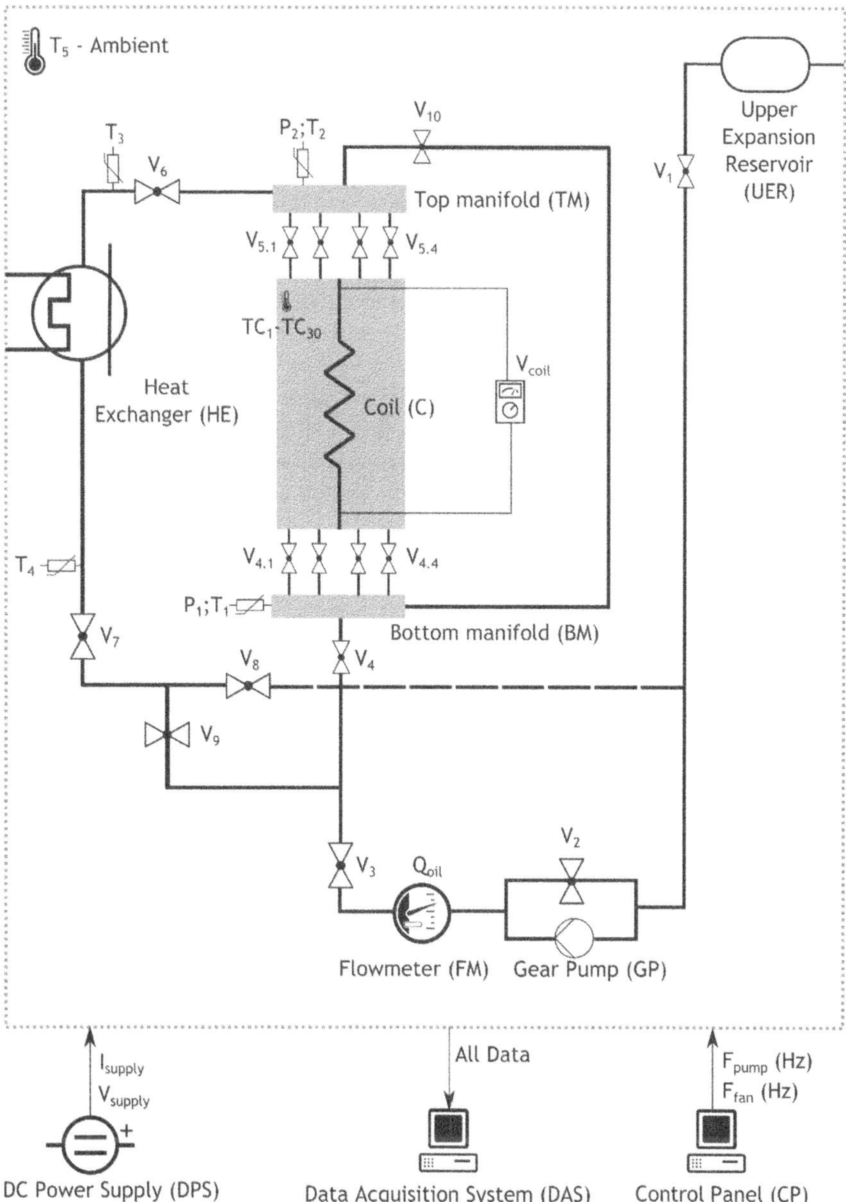

Fig. 2.4 Diagram of the experimental setup. Valves positioned to indicate the normal operation with pump

A diagram of the complete experimental setup is shown in Fig. 2.4.

The experimental setup comprises an upper expansion reservoir (UER) from which naphthenic mineral oil is fed into the coil/washer system. This expansion reservoir is directly connected to the bottom admission circuit where a gear pump (GP) guarantees the circulation of the oil through the cooling loop. The pump is equipped with a frequency controller. An ultrasonic flow meter is installed downstream the pump in order to measure the oil flow rate, Q_{oil}, before entering the bottom manifold (BM), wherein temperature and pressure sensors are located (P_1 and T_1). From this manifold, the oil is forced to flow vertically through fluid channels which are in direct contact with a heated coil (C) made of copper. In turn this heated copper coil is sandwiched between two acrylic plates where 30 thermocouples are inserted ($TC_1 - TC_{30}$). At the top of the installation, the oil exits the coil and enters a top manifold (TM) identical to the bottom counterpart. In this top manifold, additional temperature and pressure sensors are located (P_2 and T_2). From here the oil can be directly re-admitted to the bottom manifold or, under normal operation, the oil is directed towards a vertical plate radiator where additional temperature sensors are located (T_3 and T_4). A fan, also equipped with a frequency controller, is in the bottom part of the radiator. The radiator and the fan together form the Heat Exchanger (HE) identified in the diagram. After the oil flows through the vertical plates the cooling loop is closed as the oil is re-admitted to the circuit upstream the pump. The loop is equipped with several globe valves ($V_1 - V_{10}$) which are positioned in the diagram to represent the normal operation with pump. The ambient temperature is also measured using sensor T_5.

The pipe surfaces as well as the acrylic plates in direct contact with the heated coil have been insulated to minimize losses of heat to the ambient air. The heat is generated using a Direct Current (DC) power supply that imposes a DC current flow through the copper conductors of the coil. Commercial shell-type coils operate under AC, but here DC has been used to prevent additional uncertainties concerning additional eddy or stray losses.

The current and the voltage of the supply, I_{supply} and V_{supply}, are measured and can be adjusted. An additional digital multimeter is installed in the terminals of the copper coil to measure directly the voltage across the coil, V_{coil}.

The hydraulic part of the setup is controlled using a panel (CP) where the frequency of the pump, F_{pump}, and the frequency of the fan, F_{fan}, can be adjusted.

All the relevant variables mentioned ($T_1 - T_5, P_1 - P_2, TC_1 - TC_{30}, I_{supply}, V_{supply}, V_{coil}, F_{pump}, F_{fan}, Q_{oil}$) are recorded in regular time intervals of 10 s and centralized in a data acquisition system (DAS).

Next a more comprehensive description of each component of the setup is given.

2.2.2.1 Coil (C)

As shown in the detailed view of Fig. 2.3 the coil comprises an assembly of different sub-components. The coil has been manufactured using standard procedures. Figure 2.5 shows the copper coil supported in one of the acrylic plates during the assembly stage. Figure 2.5a shows the coil with all the surrounding pressboard structures except the outer insulation frame. Figure 2.5b shows the same coil after

Fig. 2.5 Coil being assembled **a** without outer insulation frame and **b** with outer insulation frame

assembling the outer insulation frame. The inner and outer insulation frames have been extended 7 mm to cover completely the innermost and outermost turns, respectively.

The detailed composition of the copper coil, including dimensions, is depicted in Fig. 2.6. Figure 2.7a depicts the complete assembly after grouping the structure shown in Fig. 2.5b together the other acrylic plate.

The total coil structure has a height of 1630 mm and a width of 1210 mm. The holes near the rubber rings represent the locations where metallic screws are inserted to maintain the structure adequately tightened. The average distance between these holes is 447 mm, while the width available for the oil flow is 271 mm (between consecutive rows of spacers). Commercial shell-type transformer windings comprise several coils piled together, and thus in these cases the coil is expected to be compressed by the weight of the neighbouring coils. The arrows indicate 4 locations through which the oil is injected (in the bottom part) and the symmetrical locations through which the hotter oil exits the coil structure towards the heat exchanger (in the top part). The oil is injected through these 4 locations in a pre-chamber, located downstream the copper coil that has been designed to

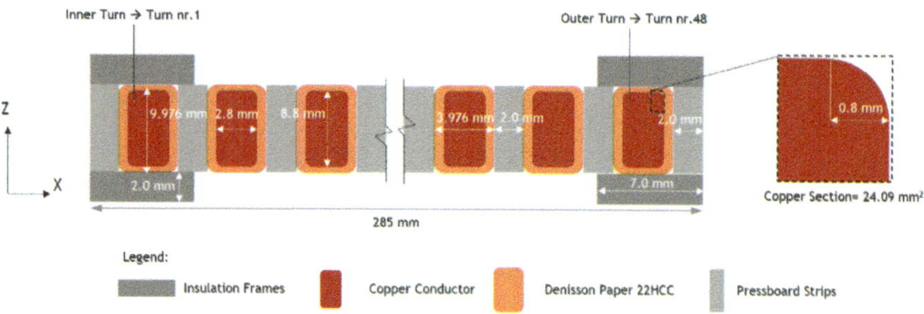

Fig. 2.6 Cut view of the copper coil with dimensions and materials

guarantee a homogeneous pressure distribution in the oil before it enters the region of main interest. This pre-chamber is intended to reproduce the bottom tank in a commercial transformer. In this bottom part of the tank, the cold oil returning from the heat exchanger suffers a sudden infinite expansion into a much larger volume of oil that is located downstream the windings. In fact, this bottom tank effect is common to both core-type and shell-type transformers, however such effect has never been comprehensively analysed and the only known reference is in the report of the CIGRE WG A2.38 (Cigre 2016), where the results seem to indicate a homogeneous pressure distribution in the bottom tank. In the scale model, this pre-chamber has an extension along the X coordinate comprised between the dashed white lines in Fig. 2.7a hence wetting the outermost turn along this extension. As this turn is completely covered by the outer insulation frame, this detail is expected to enhance the cooling conditions of this turn comparing with the innermost turn. This effect will be considered in the CFD modelling described in Chap. 3. Figures 2.7b, c intend to highlight the composition of the coil structure along the Z coordinate which has a total thickness of 53.976 mm composed by the copper coil with 9.976 mm, spacers with 2 mm thickness on both sides of the copper coil and two acrylic plates of 20 mm thickness each.

Due to the operating pressure and temperature of the fluid, the acrylic plates are expected to deform, therefore an additional reinforcing steel structure has been designed and built according to Fig. 2.8.

This steel structure has been designed to guarantee an average deformation of the acrylic plates between 0.03 and 0.05 mm for the expected range of oil flow rates under test. The maximum average deformation allowed of 0.05 mm correspond to approximately 3% of the fluid channels height, hence minimizing the impact these deformations can have by creating fluid channels with different dimensions than intended.

The blue regions shown in Fig. 2.8a correspond to polystyrene panels distributed over frontal and back acrylic plates to minimize the heat losses to the ambient air and hence guaranteeing the oil flow in the coil is as adiabatic as possible.

The orange and green cables observed at the top region guarantee the connection between the terminals of the coil and the DC power supply. These cables are made

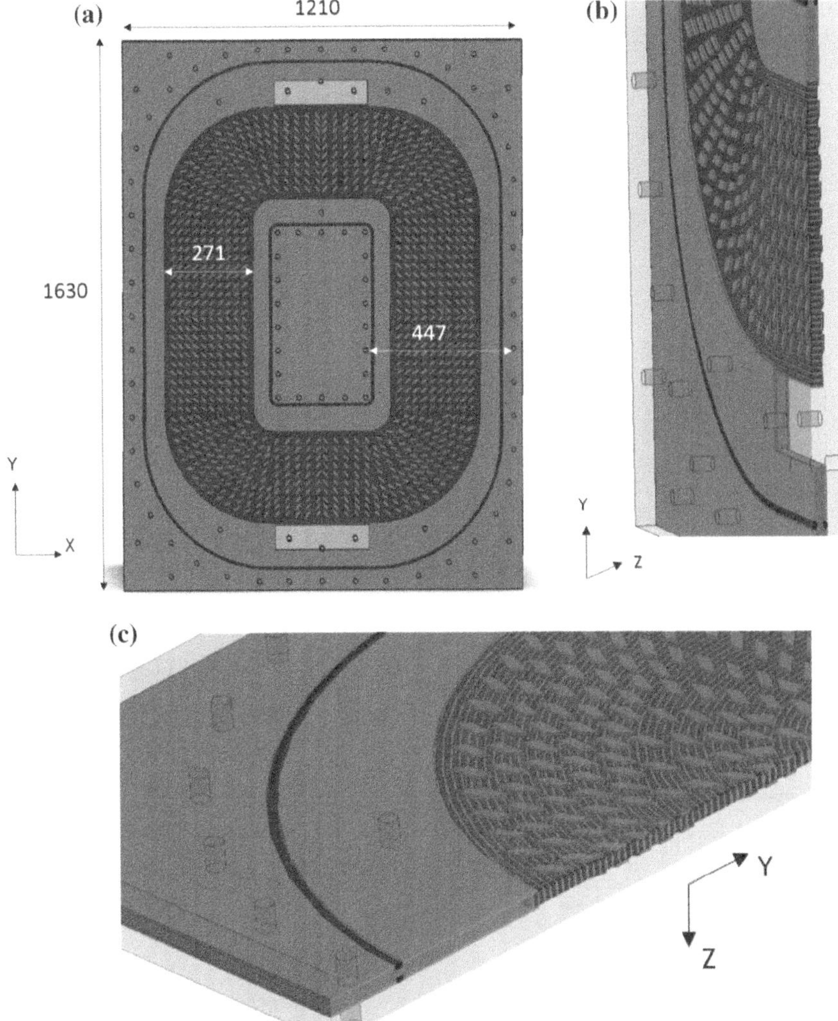

Fig. 2.7 a Coil structure with dimensions (in mm) with inlet and outlet locations identified (**b**) and **c** cut views to highlight the pre-chamber

of copper, each having a section of 90 mm^2 and a length of approximately 2 m between the terminals of the coil and the terminals of the power supply. For the highest current densities tested, these cables dissipate approximately 10% of the total power being supplied to the coil, hence an additional resistance measurement directly in the coil terminals has been used (shown in Fig. 2.9).

As mentioned above, the orange and green cables connect the coil terminals to the DC Power Supply where it is possible to measure directly the current and voltage, in every instant. Figure 2.9 shows that for a current of 144.5 A the voltage

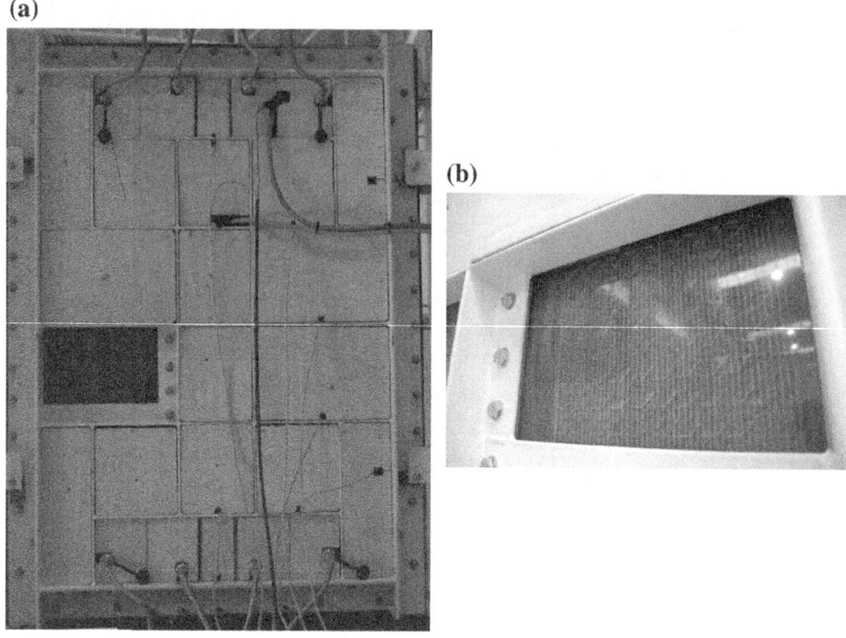

Fig. 2.8 Additional reinforcing steel structure used to minimize deformations in the coil: **a** global perspective and **b** zoomed perspective

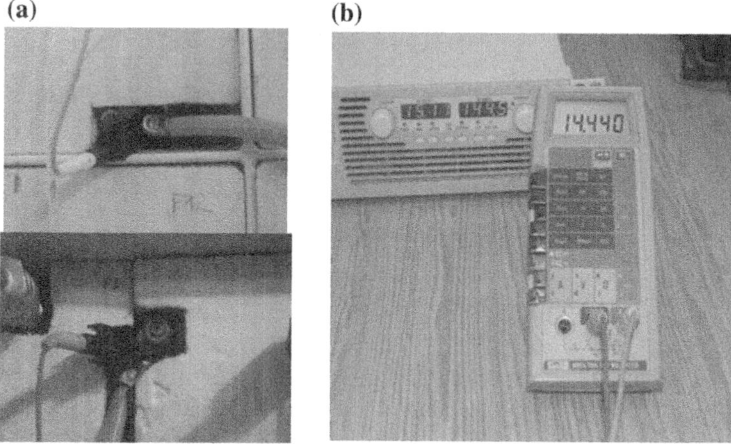

Fig. 2.9 Additional resistance measurement directly at coil terminals: **a** probes of the additional multimeter connected to the coil terminals and **b** panel of the power supply (behind) and of the multimeter (in front)

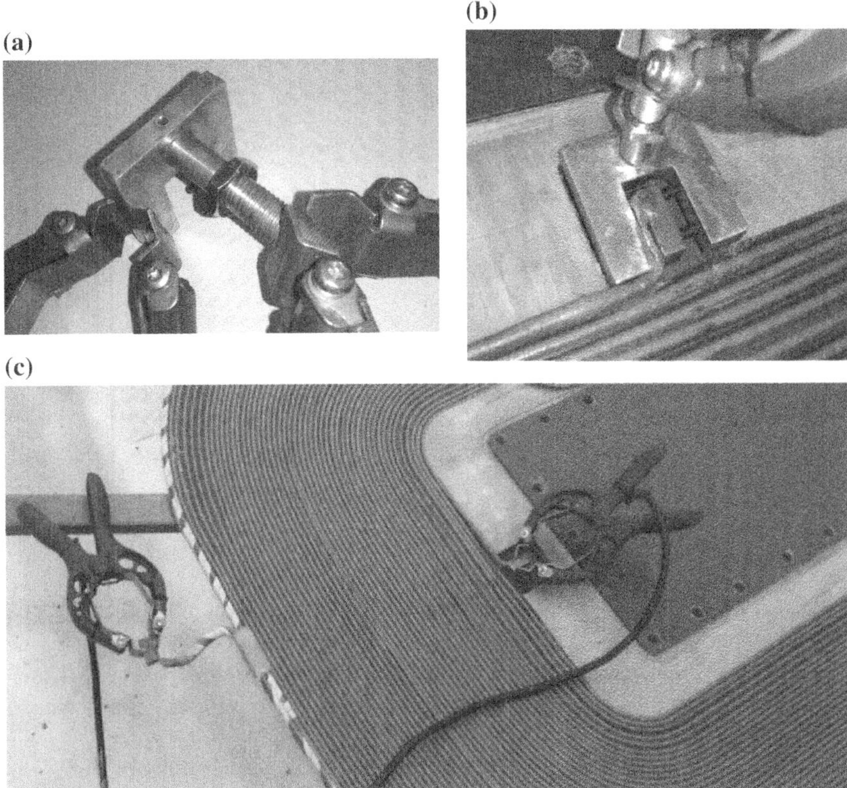

Fig. 2.10 Resistance measurements in the coil terminals: **a** individual terminal **b** terminal together with the copper coil and **c** only the copper coil

indicated in the power supply unit is 15.11 V, V_{supply}. Meanwhile it is possible to observe that a digital multimeter directly connected to the coil terminals indicates 14.44 V, V_{coil}. This voltage drop represents the cables effect and it has been adequately accounted for in every experiment conducted. This digital multimeter is a FLUKE® model 8062A and measurements have been collected manually for average time intervals of 8 min during each experiment herein described. This data is further stored in a HP xw4400® Workstation as a part of the data acquisition/control system (further details about this system are given below).

The coil terminals have been designed to guarantee two simultaneous aspects: a minimum additional ohmic resistance and low localized temperatures. For that purpose, the coil terminals have been designed as shown in Fig. 2.10a.

At room temperature, the copper coil exhibited an ohmic resistance of 101.9 mΩ. At the same room temperature, the two coil terminals introduced an additional resistance of 0.139 mΩ which accounts for a negligible amount of 0.13%

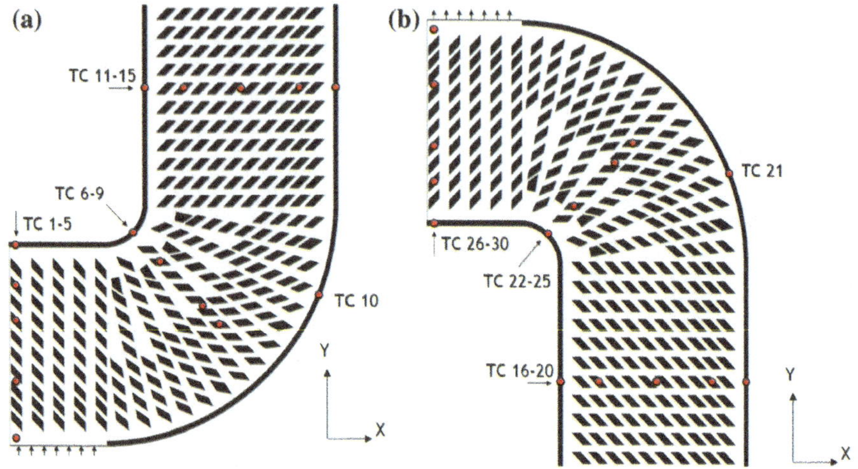

Fig. 2.11 Location of the 30 thermocouples drilled in the frontal acrylic plate (with nomenclature)

of the ohmic resistance of the copper coil. The reference ohmic resistance, $R_{coil,ref}$, for this copper coil is 100.1 mΩ at an ambient temperature of 19.6 °C.

Additionally, in order to measure the local temperatures in the coil, 30 thermocouples have been drilled in the frontal acrylic plate according to the locations shown in Fig. 2.11.

Figure 2.11a, b show that the 30 thermocouples have been distributed symmetrically over one half of the frontal acrylic plate. The sensors have been equally distributed among vertical $(TC_1 - TC_5; TC_{26} - TC_{30})$, curved $(TC_6 - TC_{10}; TC_{21} - TC_{25})$ and horizontal $(TC_{11} - TC_{20})$ regions of the coil:

- 12 thermocouples have been installed over the spacers aiming to measure copper coil temperatures $(TC_2, TC_3, TC_7, TC_9, TC_{12}, TC_{14}, TC_{17}, TC_{19}, TC_{23}, TC_{25}, TC_{27}, TC_{28})$;
- 8 thermocouples have been installed over regions where oil is expected to be flowing aiming to measure oil temperatures $(TC_4, TC_5, TC_8, TC_{13}, TC_{18}, TC_{24}, TC_{29}, TC_{30})$;
- 10 thermocouples have been installed over the insulation frames where the hottest temperatures are expected $(TC_1, TC_6, TC_{10}, TC_{11}, TC_{15}, TC_{16}, TC_{20}, TC_{21}, TC_{22}, TC_{26})$.

All the thermocouples are type-T class 1 with a systematic uncertainty of ±0.5 °C according to IEC 584.2, 1982 (BS EN 60584.2:1993). To prevent oil leakages, the thermocouples have been installed inside blind holes drilled in the frontal acrylic plate with specified depths. Figure 2.12 details how each thermocouple has been installed.

According to the above description, 8 thermocouples have been located over oil regions and 22 thermocouples over solid structures, either insulation frames or

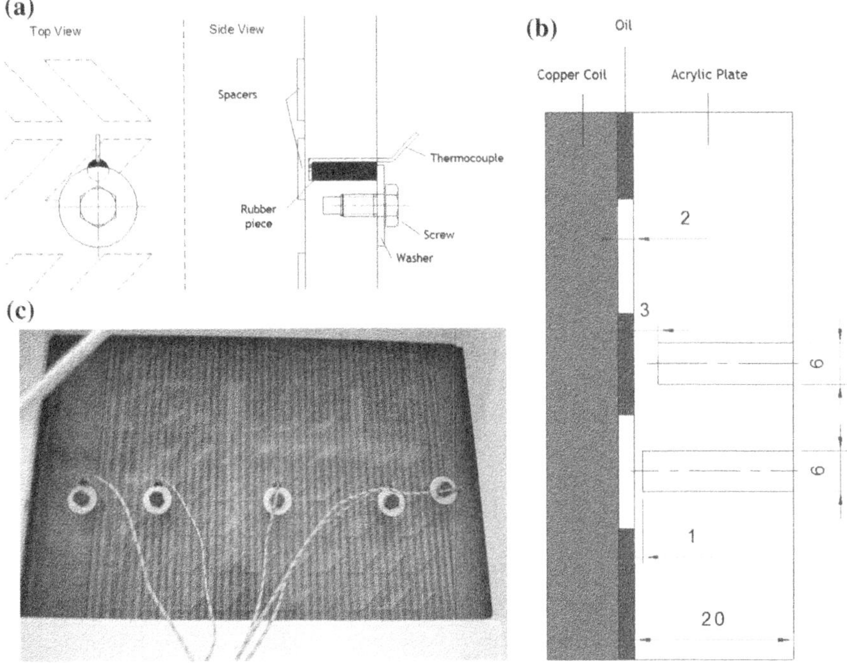

Fig. 2.12 Installation of the thermocouples in the frontal acrylic plate: **a** assembly; **b** blind hole types and dimensions and **c** photo of 5 thermocouples installed

spacers. The blind holes drilled to accommodate the 8 thermocouples aiming to measure oil have a depth of 17 mm while the blind holes aiming to measure solid temperatures have a depth of 19 mm and both are shown in Fig. 2.12b. Both blind holes are cylindrical and have a diameter of 6 mm. As the thermocouples are not in direct contact with the hot oil or the hotter copper coil surfaces, the CFD model described in Chap. 3 also includes the acrylic plates.

Each thermocouple has been installed according to the procedure depicted in Fig. 2.12a, where a rubber piece is accommodated together with the thermocouple inside each blind hole. To press the measuring tip of the thermocouples against the acrylic surface an additional washer is screwed on the top of it. The thermocouples $TC_{16} - TC_{20}$ are shown as installed in Fig. 2.12c.

The thermocouples are then connected to a YOKOGAWA® MW100 unit with 30 individual channels. Measurements have been recorded within regular time intervals of 10 s. This unit is directly connected to the HP xw4400® Workstation as a part of the data acquisition system (further details about this system are given below).

2.2.2.2   Heat Exchanger (HE)

The heat exchanger used is a vertical plate radiator with 7 plates. A fan has been installed in the bottom part of the plates to enable operating modes either with natural air flow and forced air flow. According to IEC 60076-2 standard (IEC 2011), this correspond to ODAN and ODAF cooling modes, respectively.

The radiator is shown schematically in Fig. 2.13a and the corresponding photo is shown in Fig. 2.13b. The radiator includes 7 individual plates with a height of 800 mm (along the Y coordinate) and a width of 520 mm (along the Z coordinate). The plates are separated at uniform intervals of 45 mm (along the X coordinate). The distance between the geometric center of the plates and the geometric center of the copper coil is 730 mm as indicated in Fig. 2.13a. This distance as well as the dimensions of the plates correspond to the common practice in commercial power transformers. The exception is the reduced height of each plate (800 mm) due to a lower demand in terms of evacuation capacity. Moreover, the elevation of 730 mm of the plates comparing with the copper coil enables an adequate operation of the apparatus under other cooling modes without pump—ONAF and ONAN—as defined in the same IEC standard 60076-2 (IEC 2011). However, all the experiments reported in this work have been conducted in ODAN cooling mode.

The radiator shown is from Warm® S.L. The fan under atmospheric conditions and operated at 50 Hz has a nominal air flow rate of 1460 $m^3.h^{-1}$. As referred in the

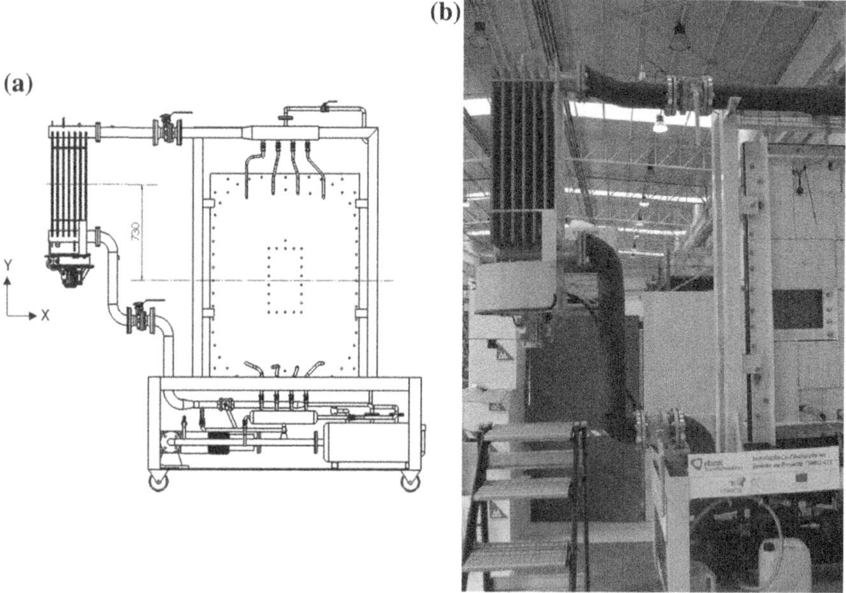

Fig. 2.13 Schematic representation of the radiators (**a**) indicating its elevation (in mm) and **b** a photo of the radiator installed with the fan below

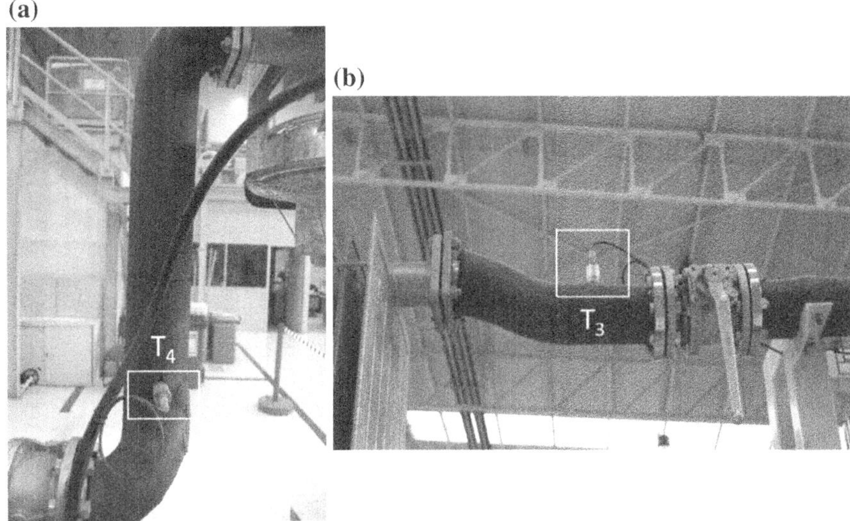

Fig. 2.14 Temperature sensors immersed in the radiators pipes: **a** upstream pipe and **b** downstream pipe

description of the whole circuit, in Fig. 2.4, the fan is equipped with a frequency controller being possible to adjust the air flow rate.

The upstream and downstream pipes of the radiators are shown in Fig. 2.14a, b, respectively. Two temperature sensors, T_3 and T_4, are in these pipes. The sensors are from IFM® model TA3130.

These are platinum resistance thermometers type PT1000 class A with a systematic uncertainty of ±0.5°C. The sensors are placed inside a metallic jacket of stainless steel with 60 mm length, which is directly immersed in the flowing oil, and hence enabling temperature measurements with improved robustness.

These temperature measurements are recorded in a centralized in control panel and further sent to a HP xw4400® Workstation as a part of the data acquisition system (further details about this system are given below). Similarly to the thermocouples installed in the frontal acrylic plate, these measurements have been recorded within regular time intervals of 10 s.

2.2.2.3 Manifolds (BM and TM)

The top and bottom manifolds indicated in the diagram of Fig. 2.4 are shown in Fig. 2.15. The manifolds are cylindrical shaped, made of stainless steel and have an approximate volume of 4 dm³ each. These manifolds intend to create a mixing chamber downstream and upstream the copper coil, enabling more homogeneous temperature and pressure measurements in these locations.

(a)

(b)

Fig. 2.15 Manifolds with sensors: **a** top manifold (with oil level indicator and air purger) and **b** bottom manifold

As identified in Fig. 2.15, pressure sensors have been installed in both top and bottom manifolds (P_2 and P_1, respectively). These are ceramic capacitive electronic sensors model IFM$^{®}$ PA3024 with a range between 0–10 bar and a systematic uncertainty of ± 0.05 bar. Additional temperature sensors T_2 and T_1 are also located in both top and bottom manifolds, respectively. These sensors are of the same model as T_3 and T_4 installed in the radiators. The objective of these specific sensors is to measure the colder fluid temperature before entering the copper coil and the hotter fluid temperature after exiting the copper coil, hence enabling the further evaluation of the enthalpy balance.

An automatic air purger has been installed in the top manifold together with an oil level indicator. This is particularly useful when the apparatus is being filled with oil by gravity.

The bottom manifold is connected to 4 flexible tubes with 17 mm of diameter. Each tube is controlled by individual valves, enabling parametric evaluations on the number of inlet sections. Even though, all the experiments reported along this thesis have been conducted with the valves of the 4 flexible tubes fully open, both in the bottom and in the top manifold ($V_{4.1} - V_{4.4}$ and $V_{5.1} - V_{5.4}$, respectively).

These temperature and pressure measurements are recorded in centralized in control panel and further sent to a HP xw4400$^{®}$ Workstation, part of the data acquisition system (further details about this system are given separately below). Similarly to the previous temperature sensors installed in the radiators, these measurements have been recorded within regular time intervals of 10 s.

2.2.2.4 Gear Pump (GP) and Flowmeter (FM)

The bottom part of the circuit identified in the diagram of Fig. 2.4 includes a gear pump and a flowmeter. Both are shown in Fig. 2.16.

The gear pump is from EFAFLU$^{®}$ having an operating point at 50 Hz of 0.5 bar and 5.4 $m^3.h^{-1}$. A frequency controller, F_{pump}, has been installed to enable adjustments in the oil flow rate, Q_{oil}, which is measured downstream of the pump using an ultrasonic flowmeter. The flowmeter model is SONOELIS$^{®}$ SE 404X with a diameter DN32. The measurement range of the flowmeter is between 0 and 20 $m^3.h^{-1}$ with a systematic uncertainty of ±5%. These flowmeter technology uses a single acoustic beam and demands a fully developed flow region to ensure adequate measurements. Therefore, straight sections of smooth DN 32 steel pipes have been used to guarantee 1000 mm upstream (>30 DNs) the device and 300 mm downstream the device (>9 DNs).

Fig. 2.16 Gear pump and ultrasonic flowmeter installed

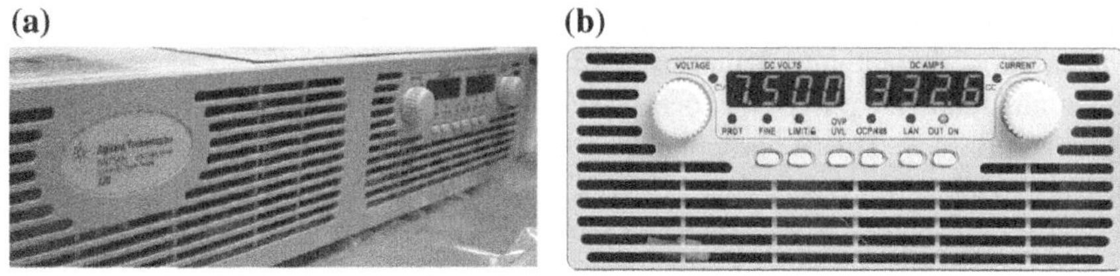

Fig. 2.17 Image of the DC Power Supply used to generate heat inside the copper coil: **a** photo and **b** schematic panel

Both devices are connected to a centralized control panel where operating variables can be manipulated, recorded and further sent to a HP xw4400® Workstation, part of the data acquisition system (further details about this system are given separately below). Similarly to the remaining sensors, these measurements have been recorded within regular time intervals of 10 s.

2.2.2.5 DC Power Supply (DCPS)

As referred in the diagram of Fig. 2.4, a DC power supply has been used to induce heat generation in the copper coil. DC current supply has been chosen in order to guarantee an exclusive Joule Effect without any additional mechanism of heat generation. The panel of this unit is shown in Fig. 2.17.

This power supply unit from Agilent® (Model N8754A) has a controllable output voltage range between 0–20 V and a controllable output current range between 0–250 A, with a total power output of 5000 W ($V_{\text{supply}}, I_{\text{supply}}, P_{total}$). The unit guarantees an output R.M.S.[1] ripple of 10 mV. This ripple has been confirmed through measurements using an oscilloscope. The systematic uncertainty of the voltage measurement is $\pm(0.025\% + 25$ mV) and the same uncertainty for the current measurement is $\pm(0.1\% + 750$ mA). These percentages correspond to the magnitude of the quantity being measured. This unit allows a continuous measurement of the ohmic resistance of the copper coil through every experiment, hence enabling an adequate indirect measurement of its average temperature. In a commercial transformer, this is not possible and the average temperature of each winding is obtained through extrapolation, after the transformer has been shut-down, according to the standard method reported in IEC 60076-2 (IEC 2011).

This unit is directly connected to a HP xw4400® Workstation, from where it can be adjusted and wherein measurements are recorded (further details about the data acquisition/control system are given separately below).

---

[1]R.M.S. is the acronym of root mean square.

2.2.2.6 Data Acquisition/Control System (DACS)

This experimental setup comprises several devices and sensors with different communication protocols and efforts have been conducted to concentrate, as much as possible, the relevant data in a single device (HP xw4400® Workstation) with the intention to minimize human intervention and generate relevant information about each experiment using objective criteria.

The diagram in Fig. 2.18 represents the data fluxes while identifying all the variables measured/controlled in this scale model during each experiment.

According to the diagram, this scale model has 4 adjustable input variables (represented by the arrows pointing inwards) and 42 output quantities (represented by the arrows pointing outwards).

The 4 adjustable input variables include:

1. The frequency of the fan and the frequency of the pump, F_{fan} and F_{pump}, both adjustable from a control panel attached to the apparatus. This control panel is shown in Fig. 2.19.
2. The current and voltage of the supply, I_{supply} and V_{supply}, both adjustable directly in the panel of the DC Power Supply Unit, shown in Fig. 2.17.

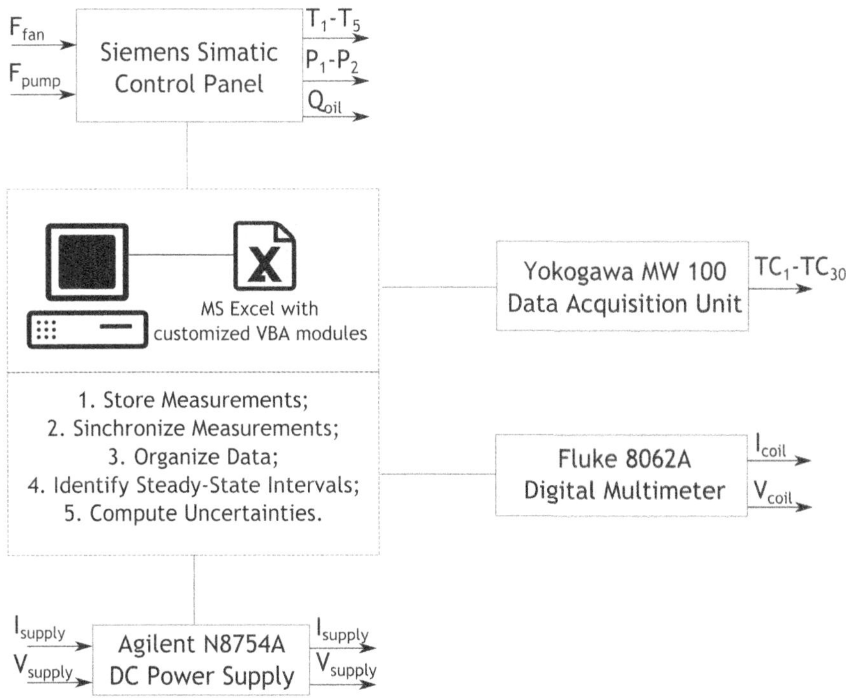

Fig. 2.18 Diagram of the data acquisition system

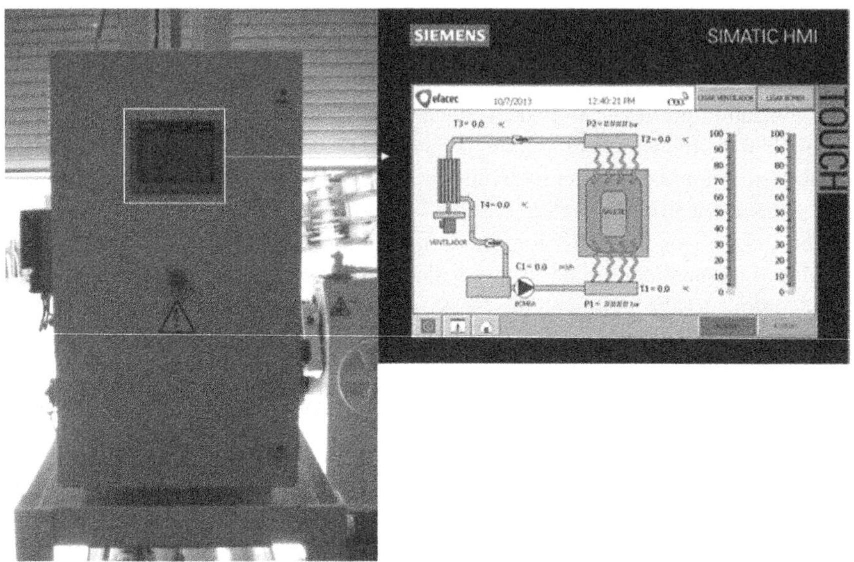

Fig. 2.19 Control Panel (CP) of the experimental setup

The 42 output quantities include:

1. Measurements of the 4 thermometers located in the bottom and top manifolds, T_1 and T_2, as well as in the radiator pipes, T_3 and T_4. An additional measurement of the ambient temperature, T_5, is also collected;
2. Measurements of the 2 pressure sensors located in both bottom and top manifolds, P_1 and P_2, respectively;
3. Measurement of the volumetric oil flow rate using the ultrasonic flowmeter, Q_{oil}.
4. Measurements of the 30 thermocouples located in the frontal acrylic plate, $TC_1 - TC_{30}$.
5. Current and voltage measurements of the copper coil collected using the digital multimeter, I_{coil} and V_{coil}.
6. Current and voltage measured using the DC Power Supply unit, I_{supply} and V_{supply}. Even though initially these values consist of inputs, these values change as the copper coil and the cables increase its temperatures.

As this data is collected from different devices, having records organized in different formats, a dedicated group of MSExcel VBA® modules has been created to automatically generate the most relevant information about each experiment.

The next section describes the methodology followed in each experiment.

2.3 Experimental Methodology

The aim of the experiments conducted is to collect measurements under steady-state conditions for comparison with results from 3D CFD and FluSHELL simulations in Chaps. 3 and 5, respectively. Some of the quantities are directly collected while other are indirectly obtained from related measurements. This section describes the methodology followed step by step.

Step 1—Filling the setup. Before each experiment or set of experiments, the first step involves filling the setup with mineral oil. The diagram of the circuit during this step is shown in Fig. 2.20 where the blue coloured lines indicate the locations filled with oil.

As can be observed in Fig. 2.20, and only for filling purposes, all the valves are opened and the pump is turned off. The circuit is filled using an upper expansion reservoir placed at the highest location of the setup and from where oil flows into the setup by the natural action of gravity.

Two conditions must be observed to move on to the next step:

- the oil level indicator shown in Fig. 2.15, shall be observed to be full of oil;
- the automatic air purger, also shown in Fig. 2.15, shall have terminated to emit its typical sound.

At this moment, the setup is full of oil and free from entrained air, namely inside the copper coil. The valves V_2, V_9 and V_{10} are then closed and the system is re-positioned back into its normal arrangement as depicted in the diagram of Fig. 2.4.

Step 2—Measuring the Average Coil Temperature. The next step is triggered when operating frequencies for the fan and for the pump are defined in the Control Panel. At this moment, a certain current is also defined in DC Power Supply unit, the copper coil starts to heat and the experiment or set of experiments is initiated.

Meanwhile, 4 values of current and the voltage start to be recorded manually at non-regular intervals of approximately 8 min:

- current and voltage values measured with the Agilent DC Power Supply Unit, I_{supply} and V_{supply}, are acquired and used to estimate a total ohmic resistance of the setup

$$R_{total} = \frac{V_{supply}}{I_{supply}} \tag{2.16}$$

- current and voltage values measured by the Fluke Digital Multimeter installed in the terminals of the copper coil, I_{coil} and V_{coil}, are also acquired and used to estimate exclusively the ohmic resistance of the copper coil:

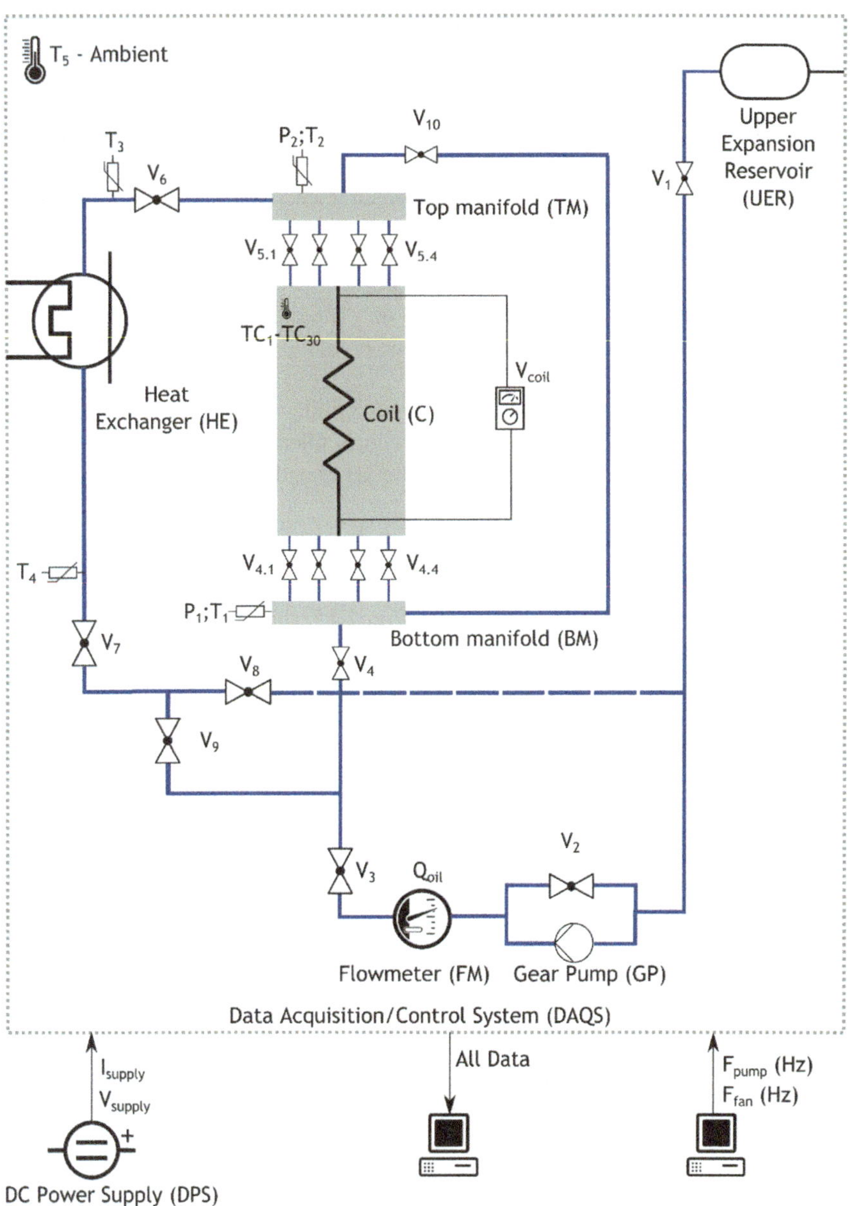

Fig. 2.20 Diagram of the circuit during the filling step

$$R_{coil} = \frac{V_{coil}}{I_{coil}} = \frac{V_{coil}}{I_{supply}} \tag{2.17}$$

With the values of the supplied current and the total ohmic resistance of the setup it is possible to calculate the total power, P_{total}, supplied to the setup according to

$$P_{total} = R_{total}I_{supply}^2 = V_{supply}I_{supply} \tag{2.18}$$

Using a similar expression, but with the values of the supplied current and the ohmic resistance of the copper coil, it is possible to calculate the total power being effectively dissipated inside the copper coil, P_{coil}:

$$P_{coil} = R_{coil}I_{supply}^2 = V_{coil}I_{supply} \tag{2.19}$$

The power dissipated in the DC cables connecting the supply unit and the copper coil can then be obtained by

$$P_{cables} = P_{total} - P_{coil} \tag{2.20}$$

Using the measured ohmic resistance of the coil in the above equation along with the reference ohmic resistance of 100.1 mΩ at 19.6 °C (as referred above), it is possible to indirectly calculate the average temperature of the copper coil, $T_{avg,coil}$, at each instant during the experiments

$$T_{avg,coil} = \frac{(235 + T_{ref})R_{coil}}{R_{coil,ref}} - 235 \tag{2.21}$$

where 235 represents a conversion constant widely used for copper conductors, as suggested in IEC 60076-2. R_{coil} represents the ohmic resistance of the copper coil measured at each instant and $R_{coil,ref}$ represents the reference ohmic resistance measured at a reference temperature, T_{ref}.

A typical evolution of the average temperature of the copper coil, $T_{avg,coil}$, is shown in Fig. 2.21 for a set of three experiments conducted sequentially.

Figure 2.21 shows an initial longer transient when the whole system is heating up. Then, after approximately 473 min the copper coil stabilized at 78.5 °C, with a current of 153 A. After 523 min a new current has been set in the DC power supply unit and the copper coil went through another transient up to another stabilization period, after 690 min, at 66.0 °C with a current of 134 A. After 752 min a new current has been set. The copper coil went through another transient and finally stabilized, after 929 min, at 52.9 °C with a final current of 112A.

The first steady-state period at 78.5 °C has been achieved after approximately 7.9 h and this time constant correspond to the time needed for the all components to

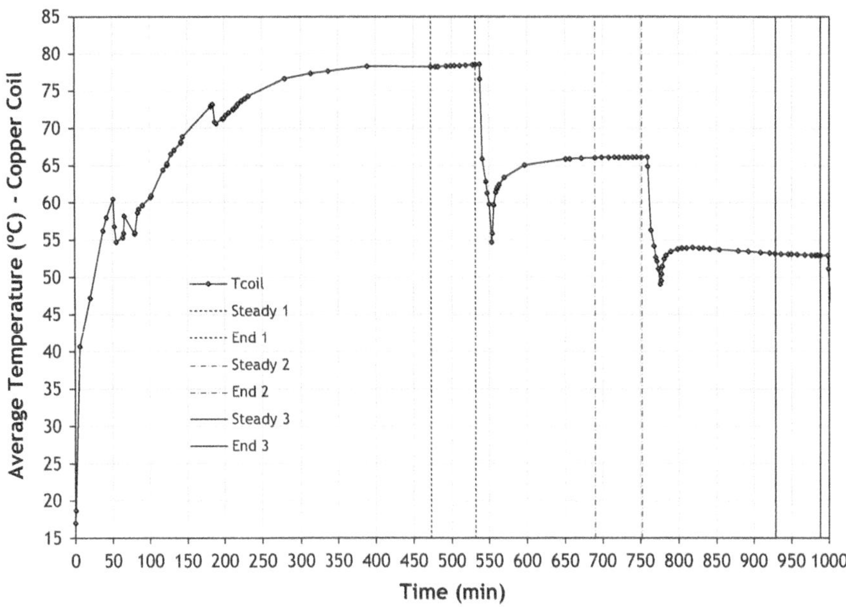

Fig. 2.21 Average coil temperature evolution over a set of three consecutive experiments (three steady-state intervals identified)

heat up. Not only the copper coil but also other components such as the oil, the acrylic plates, the pressboard structures as well as the pipes.

Some of the peaks observed during the initial transient correspond to higher currents and low oil flow rates that have been imposed in the meantime to accelerate the time needed for the system to stabilize. As the aim of these experiments is to collect information under steady-state conditions, the regions of interest correspond exclusively to these stabilization periods. Consequently, the curve depicted in Fig. 2.21 refer to a set of three experiments corresponding to three steady-states (Steady 1, Steady 2 and Steady 3) conducted sequentially where the only variable differing from experiment to experiment is the final DC current imposed and hence the corresponding losses being dissipated inside the copper due to Joule effect.

This procedure is simpler than the temperature rise tests usually conducted in commercial power transformers to assess whether it is conformal or not with the guaranteed temperature limits. These tests usually held in the HV Laboratories of each manufacturer before a unit is commissioned and guidelines for this are standardized in IEEE C57.12.90 (Board 2006) and IEC 60076-2 (IEC 2011).

One of the major differences is that commercial power transformers operate under AC conditions, so the electrical circuit involved is more complex. In that cases the transformers need to be de-energized before a group of DC batteries can be connected to the terminals of each main winding to measure the decayment of the ohmic resistance over a typical period of 30 min. However, there is an intrinsic time gap between the instant when the transformer is de-energized and the instant when the ohmic resistance decayment starts to be measured, which is typically between 2 and 4 min. Consequently, the ohmic resistance at the exact instant when the transformer is de-energized needs to be extrapolated according the dynamics of the system observed in the decayment curve. An additional challenge to this procedure, is that the time constant of the copper windings is much lower when compared with other components of the transformer, such as the larger oil volume. Again some guidelines for this extrapolation are standardized IEC 60076-2, more specifically in the informative Annex C (IEC 2011).

In this experimental setup, and according to the procedure described, these limitations are not present as the ohmic resistance of the copper coil can be calculated at every instant during the experiments using the current and voltage measured directly at the terminals of the coil, I_{coil} and V_{coil}.

In this experimental setup, the average copper coil temperature is considered to have achieved its steady-state temperature when its temperature reached a variation lower than $1\ °C.h^{-1}$ and remained there for a period of 1 h. The temperature variations are not calculated based on the previous value recorded, but on the value recorded 1 h before which implies at least 1 h of runtime before start calculating variations and at least 1 h after the criteria has been observed for first time. Afterwards, the average copper coil temperature is calculated averaging the last hour of measurements under steady-state conditions. Furthermore, the temperatures measured during this 1 h steady-state period are further used to compute a random uncertainty associated. A more detailed description on the computation of uncertainties is given separately below.

For commercial transformers the stability criteria defined both in IEC60076-2 (IEC 2011) and in IEEE C57.12.90 (Board 2006) is a temperature variation of the top oil temperature rise below $1\ °C.h^{-1}$ over consecutive period of 3 h. The way the temperature variations must be calculated is not clear in the standards, but the lower relative oil volumes involved in this setup, justify the current stability criteria adopted.

Step 3—Measuring Local Acrylic Temperatures, Oil Temperatures, Flow Rate and Pressures. According to the diagram shown in Fig. 2.18, another 38 output quantities begin to be stored in regular time intervals of 10 s. As mentioned before, 8 of the quantities are acquired via the SIMATIC® Control Panel and the remaining 30 are acquired via an independent unit from YOKOGAWA®.

The data is acquired in different formats, so in order organize the data, a dedicated MSExcel® environment has been customized for this purpose. This environment is shown in Fig. 2.22.

In this environment, some VBA® modules have been customized to automatically:

- Store the measurements compiled in a single environment.

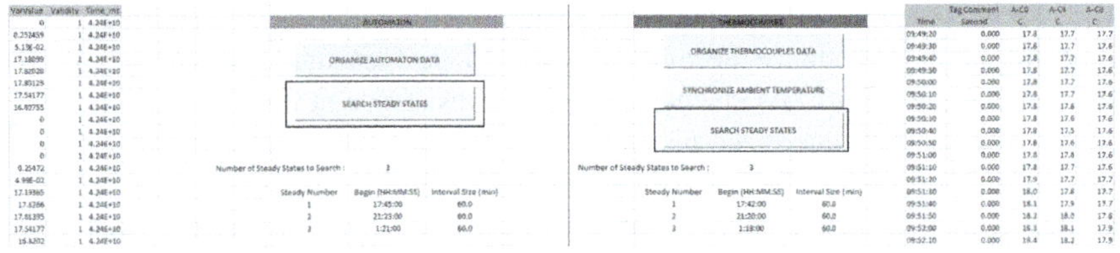

Fig. 2.22 Customized MSExcel® environment developed to systematize the data collected

- Synchronize all the measurements, hence guaranteeing that for every instant a certain oil temperature (e.g. T_1) has a correspondent temperature in the same temperature in the acrylic plate (e.g. TC_1).
- Organize all 38 output quantities within the same format.
- Enable an objective identification of the steady-state intervals complying with a pre-defined stability criteria. This is probably the most relevant characteristic to guarantee a reproducible analysis of each experiment without subjective considerations.
- Compute the combined uncertainties associated with each measurement.

Figure 2.23 shows the evolution of ambient and oil temperatures for the same set of three consecutive experiments for which the average copper coil temperature has been depicted in Fig. 2.21.

The ambient temperature did not present abrupt variations during the experiments, which have been held in a closed room. Nevertheless, the ambient temperature has been synchronized and subtracted to every temperature measurement to avoid this influence. This means that every temperature measurement has been transformed into its correspondent temperature rise above ambient, which is a

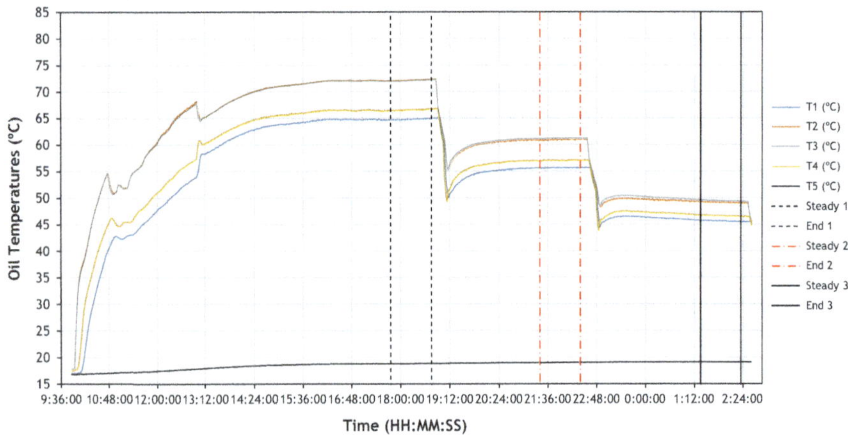

Fig. 2.23 Oil temperature evolution over a set of three consecutive experiments (three steady-state intervals identified)

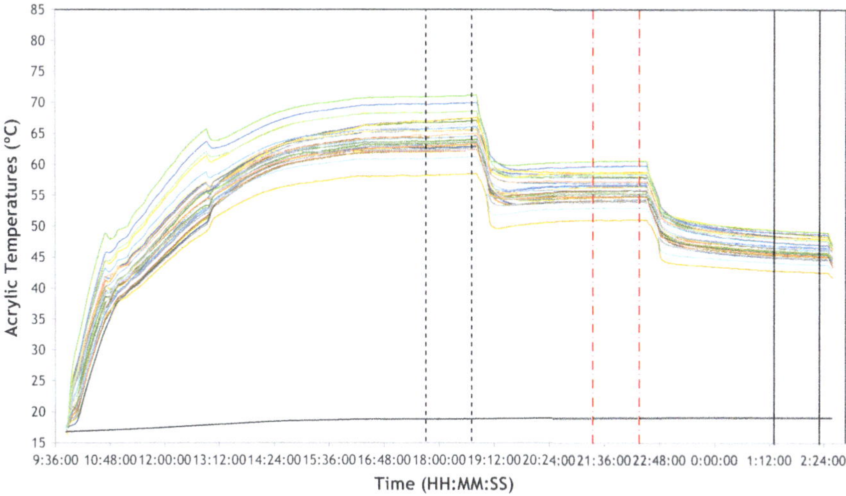

Fig. 2.24 Acrylic temperatures evolution over a set of three consecutive experiments (three steady-state intervals identified)

concept widely used in transformer industry and standardized both in IEC 60076-2 (IEC 2011) and IEEE C57.12.90 (Board 2006).

The different sensors might have different response times, so the stability criterion is then applied simultaneously to the 35 temperature rise measurements (5 oil temperature rises and 30 acrylic temperature rises) as well as to the average copper coil temperature rise described earlier.

The slowest measurement observing the stability criterion defines the starting time of each steady-state interval. For instance, in a first experiment when the system is heating up from a colder temperature, the acrylic temperatures have been observed to achieve the steady-state later than the oil temperatures. The acrylic temperatures evolution is shown in Fig. 2.24.

Finally, the last hour of measurements is collected and used to compute the combined uncertainties.

Additional oil flow rate measurements are also collected and organized using the same MSExcel® environment. The plot in Fig. 2.25 comprises an example and reports to the same set of three previous consecutive experiments.

As shown in Fig. 2.25 the same steady-state intervals are assumed for the oil flow rate measurements. The signal oscillations have been observed during all the experiment and seem to increase when the Reynolds number in the pipes of the bottom admission circuit is below 1000. For these regimes, the buoyancy forces are dominant hence a more heterogeneous temperature distribution inside the piping is expected. As the sound propagation is temperature dependent, this seems to create additional oscillations in the flowmeter signal.

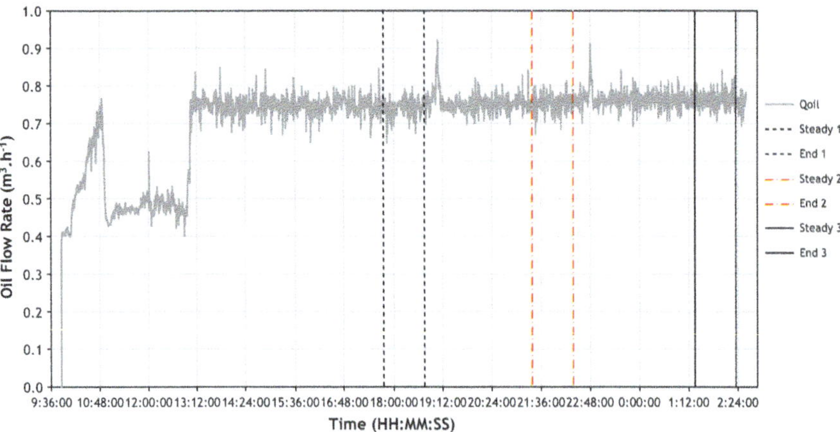

Fig. 2.25 Oil Flow rate evolution over a set of three consecutive experiments (three steady-state intervals identified)

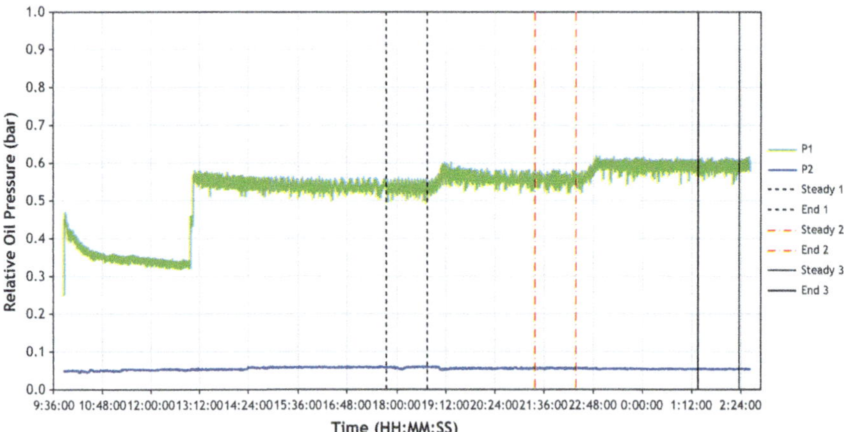

Fig. 2.26 Relative oil pressure evolution over a set of three consecutive experiments (three steady-state intervals identified)

Pressure measurements are also collected and organized using the same MSExcel® environment. Figure 2.26 depicts the evolution of the relative oil pressure measured in the bottom and top manifolds, P_1 and P_2, respectively.

The same steady-state intervals, previously identified when applying the stability criteria to all the temperature measurements, have been considered to extract the steady-state values of both pressures—as shown in Fig. 2.26.

The next step, also conducted under the same customized MSExcel$^{®}$ environment, refers to the way uncertainties are computed for each one of these measurements described in Steps 2 and 3 of the experimental methodology.

Step 4—Computing Uncertainties. Every measurement has an uncertainty associated with it and the final step of this methodology includes the computation of the total (or combined) uncertainties associated with each quantity that is to be further compared with the numerical results from CFD.

The standard guidelines for the calculation of uncertainties affecting the results of temperature rise tests in commercial transformers are described in Sect. 7.11 of the IEC 60076-2 (IEC 2011). These guidelines suggest reporting uncertainties with a confidence level of 95%.

Under the present methodology, each single measurement has been considered to have a total (combined) uncertainty related with both systematic and random features:

- Systematic features are intrinsically associated with each sensor. To account for this, a systematic uncertainty, P_x, is associated with each measurement based on the accuracy referred in the specification sheet of the correspondent sensor.
- Random features derive from the specific usage of the correspondent sensor and are statistically inferred from the measurements collected during the steady-state intervals of each experiment.

During these steady-state intervals of 1 h, measurements are stored every 10 s resulting in samples with 359 degrees of freedom. First the arithmetic mean, $\bar{x}$, and the standard deviation, σ, of these samples are calculated.

Then, as the number of degrees of freedom is high, each measurement has been assumed to obey a normal statistical distribution.

This assumption implies that 95.6% of the measurements are comprised in the interval $[\bar{x} - 2\sigma \, ; \, \bar{x} + 2\sigma]$. Hence a random uncertainty, B_x, for each measurement is calculated by

$$B_x = \pm 2\sigma \qquad (2.22)$$

Finally, both systematic and random uncertainties, are combined in a total uncertainty, U_x, using the following expression:

$$U_x = \sqrt{P_x^2 + B_x^2} \qquad (2.23)$$

The experimental measurements reported in this work are always reported together with this total uncertainty associated $\bar{x} \pm U_x$.

The average copper coil temperature, $T_{avg,coil}$, comprises a particular exception, because this quantity results from a calculation involving two measurements, V_{coil} and I_{supply}, Eq. (2.21). The reference temperature and resistance have been measured once, so they are treated as constants.

In this case the total uncertainty, $U_{T_{avg,coil}}$, results from the propagation of the combined uncertainties associated with both V_{coil} and I_{supply} as

$$U_{T_{avg,coil}} = \sqrt{\left(\frac{\partial T_{avg,coil}}{\partial V_{coil}}\right)^2 \cdot U_{Vcoil}^2 + \left(\frac{\partial T_{avg,coil}}{\partial I_{supply}}\right)^2 \cdot U_{Isupply}^2} \qquad (2.24)$$

The average copper coil temperatures further used for comparison with CFD results are consistently reported as $T_{avg,coil} \pm U_{T_{avg,coil}}$.

2.4 Conclusions

On one hand, the progressive use of more complex tools to predict the thermal behaviour of power transformers empowers the users with enhanced capabilities of directly manipulating new parameters and new details. On the other hand, the experimental validation of these tools also demand for highly-controlled experimental environments. At least, in early stages where physical assumptions need to be conceptually validated.

The current chapter describes the development of a dedicated experimental setup which reproduces a representative portion of the closed cooling-loop of a commercial shell-type power transformer—the coil/washer system. This closed cooling loop comprises a single copper coil scaled down by a factor of one third and vertical plate radiators to remove the heat generated to the ambient air. The scaled down characteristics of this setup eases a tighter control of certain parameters, minimizes the combined uncertainty associated with some measurements (namely when comparing with acceptance tests) and enables measurements that are typically not possible in commercial transformers. According to CIGRE Working Group A2.38 brochure, and to literature in general, this chapter describes the development of the most complete experimental setup for shell-type transformers built so far, involving simultaneously flow and heat transfer.

The setup has been designed to ensure representativeness of the full-scale conditions and it has been used to conduct 9 experiments, reproducing 9 combinations of realistic operating conditions.

Each experiment has been conducted with a pump to impose the oil circulation and until steady-state conditions have been observed. These experiments, correspond to a typical OD cooling regime in a commercial shell-type transformer. Variations of less than 1 $^{\circ}$C.h^{-1} have been assumed as the required criterion to consider steady-state, being this in line with the common practice during acceptance tests of commercial power transformers. After reaching steady-state, the registration of measurements has been extended during 1 h in order to estimate combined uncertainties associated with each measurement. Steady-state conditions have been reached in every experiment.

These experiments are further used to validate the corresponding CFD and FluSHELL simulations in Chaps. 3 and 5, respectively.

References

Board, I. -S. S. (2006). *C57.12.90: IEEE standard test code for liquid-immersed distribution, power, and regulating transformers.*

Campelo, H. M. R. (2015). Cooling of large power transformers using CFD: vision, strategy and case studies. In *ANSYS convergence conference, Porto.*

Cigre. (2016). *WG A2.38 Brochure, transformers thermal modelling. Draft Version 5.*

Gomes, P. J., Sousa, R. G., Dias, M. M., & Lopes, J. C. B. (2007). Large power transformer cooling—flow simulation and PIV analysis in an experimental prototype. In *Advanced research workshop on transformers, Baiona* (pp. 113–129).

Higgins, W. F., & Davis, A. H. (1928). Thermal transference in transformer oils. *Journal of the Institution of Electrical Engineers, 67,* 527–537. https://doi.org/10.1049/jiee-1.1929.0041.

IEC. (2011). *IEC60076: Power transformers–Part 2: Temperature rise for liquid-immersed transformers.*

Picher, P., Torriano, F., Chaaban, M., Gravel, S., Rajotte, C., & Girard, B. (2010). Optimization of transformer overload using advanced thermal modelling. In *Cigré, Paris* (p. A2_305_2010).

Szpiro, O., Allen, P. H. G., & Richards, C. W. (1982). Coolant distribution in disc type winding horizontal ducts and its influence on coil temperatures. In U. Grigul, E. Hahne, K. Stephan, & J. Straub (Eds.), *Seventh international heat transfer conference, Munchen, Germany.*

Taylor, E. D., Berger, B., & Western, B. E. (1957). An experimental approach to the cooling of transformer coils by natural convection. *Proceedings of the IEE-Part A Power Engineering, 105,* 141–152. https://doi.org/10.1049/pi-a.1958.0034.

Torriano, F., Campelo, H. M. R., Labbé, P., Quintela, M. A., & Picher, P. (2016). Experimental and numerical thermofluid study of a disc-type transformer winding scale model. In *XXIInd international conference on eletrical machines, Lauzanne, Switzerland.*

Weinlader, A., & Tenbohlen, S. (2009). Thermal-hydraulic investigation of transformer windings by CFD-modelling and measurements. In *Proceedings of the 16th International Symposium on High Voltage Engineering, Johannesburg, South Africa.*

Weinlader, A., Wu, W., Tenbohlen, S., & Wang, Z. (2012). Prediction of the oil flow distribution in oil-immersed transformer windings by network modelling and computational fluid dynamics. *IET Electric Power Applications, 6,* 82–90. https://doi.org/10.1049/iet-epa.2011.0122.

Wikipedia. (2016). Standard deviation [WWW Document]. http://en.wikipedia.org/wiki/Standard_deviation.

Zhang, J., Li, X., & Vance, M. (2008). Experiments and modeling of heat transfer in oil transformer winding with zigzag cooling ducts. *Applied Thermal Engineering, 28,* 36–48. https://doi.org/10.1016/j.applthermaleng.2007.02.012.

Chapter 3
CFD Scale Model

In Sect. 3.1, the most significant components of the entire experimental setup have been modelled in CFD. This model is more detailed than the CFD model described in Chap. 5 and aims to be directly compared with the experiments conducted previously described in Chap. 2. In this section the geometry of this detailed CFD model, its mesh and boundary conditions are presented together with some results.

In Sect. 3.2 a group of 9 CFD simulations is comprehensively compared against 9 experiments. The set of experiments include different oil flow velocities (between 10.3 and 19.2 cm.s^{-1}) and different heat fluxes (12.0 upto 25.4 W.dm^{-2}). The CFD predictions are compared against the measured average temperatures of the coil but also with the local temperatures collected using 30 thermocouples distributed over the whole frontal acrylic plate.

Finally, in Sect. 3.3, specific conclusions are set about whether the CFD simulations are considered validated or not.

3.1 CFD

The main objective of this work is the development of a thermal-hydraulic network algorithm capable of predicting accurately the steady-state temperature distribution in shell-type coils (this algorithm is part of the FluSHELL tool described in Chap. 4). However, the accuracy of that algorithm is dependent on correlational data extracted from CFD, the friction and the heat transfer coefficients. Thus, it is of paramount importance to validate the CFD results.

© Springer International Publishing AG 2018

65

H. Campelo, *FluSHELL – A Tool for Thermal Modelling and Simulation of Windings for Large Shell-Type Power Transformers*, Springer Theses, https://doi.org/10.1007/978-3-319-72703-5_3

3.1.1 Geometry

For this reason, a specific detailed three-dimensional CFD model of the experimental setup described in Chap. 2 has been developed to be further compared with experiments. The geometry of the CFD model is shown in Figs. 3.1 and 3.2.

The computational domain has been truncated at the surface of the bottom inlets and at the surfaces of the top outlets. In this way, the simulation of the complete cooling loop has been avoided. In fact there is no specific demand for such complete cooling simulation since the measured oil flow rate and the measured bottom oil temperatures are boundary conditions imposed in each CFD simulation.

Notwithstanding, this detailed CFD model is intrinsically different from the CFD model described in Chap. 5. That model, used to validate the FLUSHELL tool, has been intentionally simplified to reproduce the exact physical assumptions of the tool. In that model, the bottom and top oil pools have not been modelled neither the surrounding solid structures (e.g. pressboard blocks, acrylic plates). Thus, in that

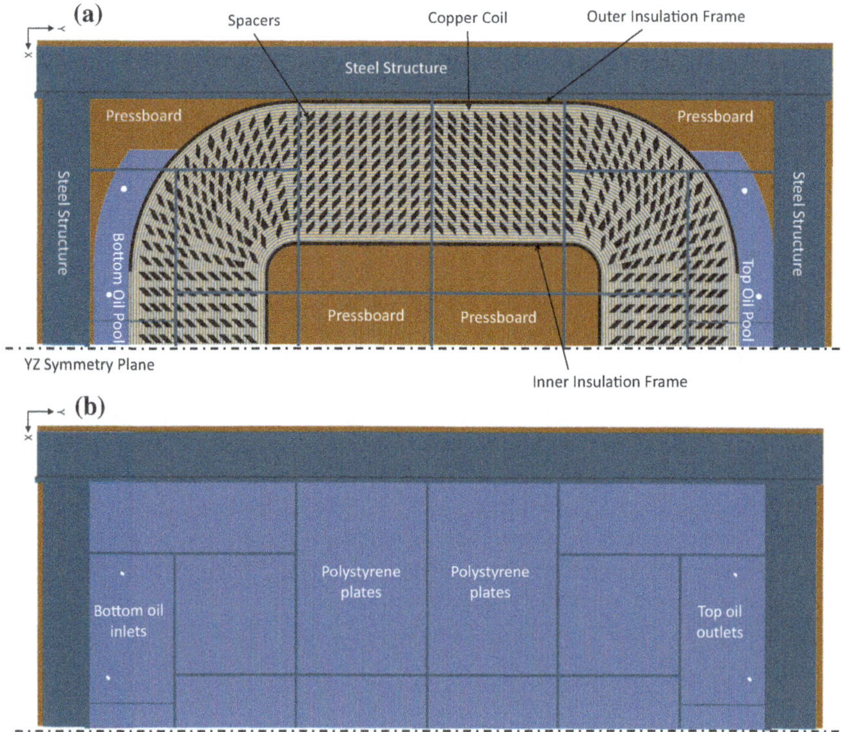

Fig. 3.1 XY view of the CFD geometry used to represent the experimental setup: **a** without the polystyrene plates and transparency on the acrylic plate and **b** with the polystyrene plates

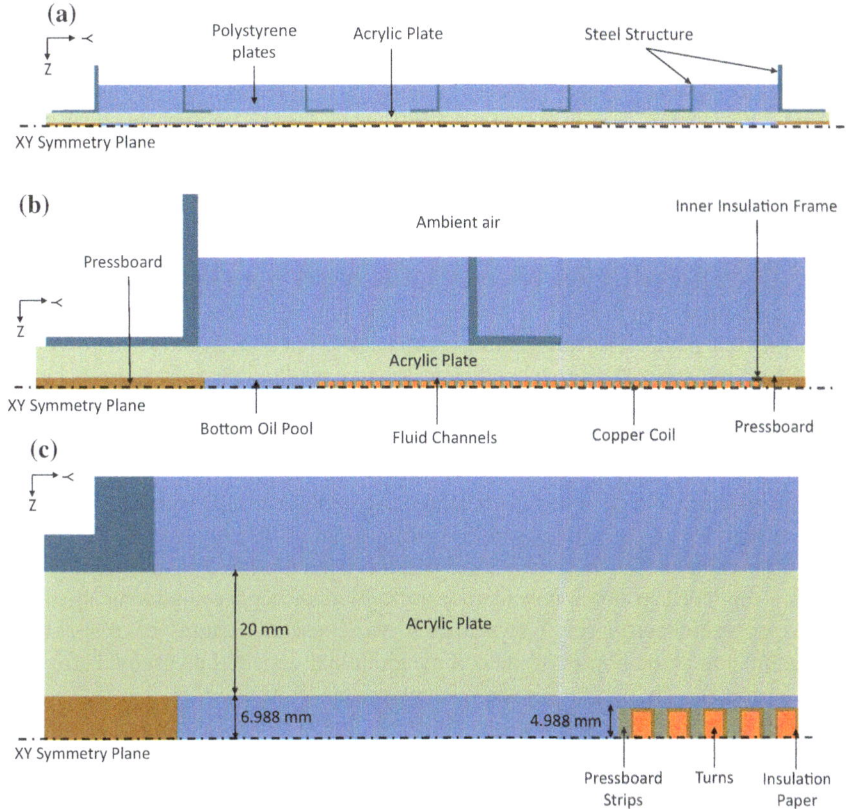

Fig. 3.2 YZ view of the CFD geometry built to represent the scale model: **a** main components along Z direction and **b** with further detail about specific components and dimensions

CFD model all those interfaces have considered adiabatic to respect the same assumptions made in FluSHELL.

However, as shown in Fig. 3.1, this CFD model aims to be compared with experiments from the scale model described in Chap. 2, thus the solid structures surrounding the coil have also been considered. The list includes:

- pressboard blocks around inner and outer insulation structures;
- the bottom and top oil pools;
- the acrylic plates;
- the external steel structures used for reinforcement;
- the polystyrene plates.

These solid structures surrounding the coil are relevant:

- on one hand, because the heated copper coil of the experimental setup exchanges heat with all these surfaces;

- on the other hand, because the thermocouples used for local temperature measurements are not in direct contact with the coil, which requires to consider the acrylic plates in the CFD model.

This CFD domain has two symmetrical planes: the *YZ* Symmetry Plane indicated in Fig. 3.1 that halves the whole domain needed and the *XY* Symmetry Plane indicated in Fig. 3.2 that divides the copper coil in two halves. Hence the final CFD domain used corresponds to one quarter of the experimental setup previously presented in Chap. 2.

Figure 3.2 shows a cut view of the CFD domain with increasing detail towards the copper coil. As shown in Fig. 3.2b the domain is limited by the external surfaces of the polystyrene plates. The surrounding ambient air is not directly modelled, instead a typical heat transfer convective coefficient for free air flows has been used (10 W. $m^{-2}.°C^{-1}$). The sensitivity of the CFD predictions to the numerical value of this coefficient have been analysed (in a range between 2–18 $W.m^{-2}.°C^{-1}$) and no significant differences have been observed. The thickness of the polystyrene plates (50 mm) has been overdesigned to guarantee a maximum heat dissipation to ambient air lower than 5% of the total power being dissipated in the copper coil.

As observable in Fig. 3.2c the copper coil has been modelled in detail. Each turn has been modelled in detail considering an individual copper conductor insulated with paper. Pressboard strips between each two consecutive turns have also been considered, and at the border of the coil an additional unpaired pressboard strip has been considered. The whole copper coil comprises 48 turns.

In terms of dimensions, materials and mesh, the coil/washer domain is exactly the same used in the CFD simulations described in Chap. 5. The other components of the model (e.g. pressboard blocks, oil pools, steel structure) have the exact same dimensions as described in Chap. 2 where the scale model has been presented. All the dimensions correspond to nominal dimensions assessed in the design stage. No adjustments to the nominal dimensions have been considered.

3.1.2 Mesh

An unstructured mesh of 32.7 Million elements has been created. A *YZ* cut view of this mesh is shown in Fig. 3.3.

Due to the complexity and the number of components requiring to be meshed, this discretization has been accomplished using several interfaces. As observed in Fig. 3.3a the mesh used in the external steel structure is not conformal with the mesh used in the acrylic plate. Similarly, it can be observed in Fig. 3.3c that the mesh used in the fluid channels is not conformal neither with the acrylic plate nor conformal with the mesh used in the turns.

The cut view of the mesh is shown Fig. 3.4. In the turns and in the fluid channels the mesh is exactly the same as used in the simpler CFD model described in Chap. 5.

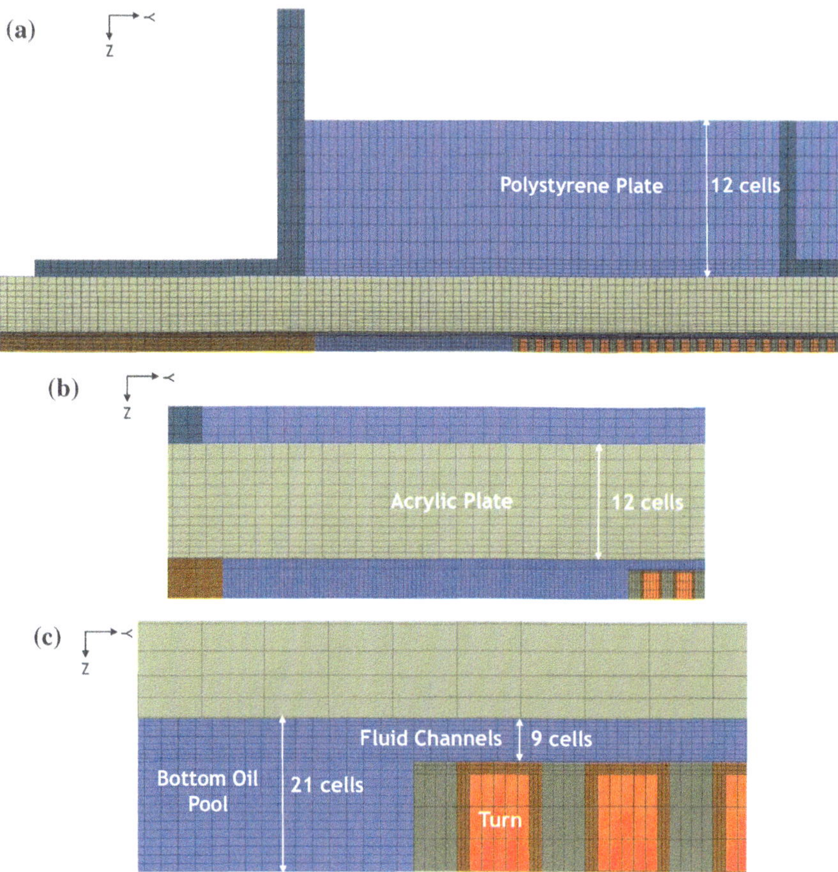

Fig. 3.3 Type of mesh elements and mesh resolution used along Z-coordinate: **a** in the polystyrene plates, **b** in the acrylic plate and **c** in the fluid channels and in the turns of the copper coil

Table 3.1 summarizes the mesh elements in each component of the current mesh. The current mesh is shown grey-shaded and compared side by side with the other mesh created for that simpler CFD model used to validate the FluSHELL tool in Chap. 5.

The current mesh has an increased number of elements: 32.7 Million elements against 25.6 Million elements. This additional number of mesh elements is due to the extend fluid domain which in this case includes the bottom and top oil pools (16.1 against 13.1) and due to the other solid structures referred above (and identified as NA in Table 3.1).

The mesh size differs from component to component but the most sensitive direction is along the Z coordinate. This direction is perpendicular to the most relevant fluid velocity and temperature profiles near the surface of the copper coil.

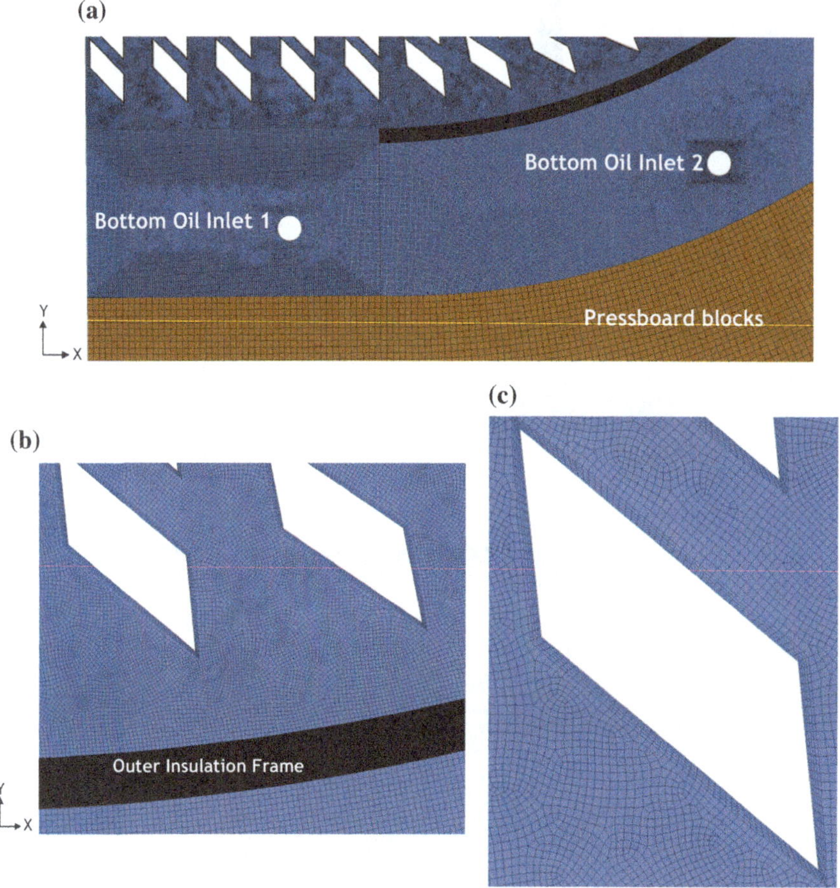

Fig. 3.4 Type of mesh elements and mesh resolution: **a** near the bottom oil inlets and **b** near the outer insulation frame and **c** around the spacers

The mesh resolution influence along this direction is assessed and discussed in Chap. 4 during FluSHELL calibration. There it is concluded that 9 elements along the Z direction is sufficient to achieve numerical uncertainties below 0.5 °C. As a result, that same mesh resolution has been used in the fluid channels of this model.

3.1.3 Boundary Conditions

A group of 9 CFD simulations has been conducted to compare with the 9 experiments conducted in this experimental setup. The boundary conditions used in these 9 CFD simulations are listed in Table 3.2. The simulations have been conducted using ANSYS Fluent® 16.2.

Table 3.1 Distribution of the mesh elements between the different components of the domain. Current CFD model versus CFD model described in Chap. 5

Component	CFD Model for FluSHELL tool validation		CFD Model to compare with experiments	
	MElements	%	MElements	%
Fluid	13.1	54.5	16.1	49.4
Spacers	6.1	24.0	6.1	18.8
Copper coil Turns	5.1	20.0	5.1	15.6
Insulation frames	0.4	0.5	1.7	5.3
Pressboard blocks	NA	NA	1.1	3.4
Acrylic plate	NA	NA	1.4	4.2
Polystyrene plates	NA	NA	0.8	2.5
Steel structure	NA	NA	0.3	0.9
Total	25.6	100	32.7	100

Table 3.2 Boundary conditions and most relevant solver parameters

Bottom oil inlets	Inlet velocity	Variable and uniform. Table 3.3.
	Temperature	Variable. Table 3.3.
Top oil outlets	Static pressure	0 Pa
Copper conductors	Volumetric heat source	Variable $W.m^{-3}$. Table 3.3.
	Momentum	No-Slip Condition
Spacers	Volumetric heat source	$0\ W.m^{-3}$
	Momentum	No-Slip Condition
Insulation frames	Volumetric heat source	$0\ W.m^{-3}$
	Momentum	No-Slip Condition
Polysterene plates	Momentum	No-Slip Condition
	Heat transfer	$10\ W.m^{-2}.°C^{-1}$ with variable ambient temperatures. Table 3.3.
Symmetries	XY plane at middle height of the turns according to Fig. 3.2	
	YZ plane in Fig. 3.1	
Solver	Pressure based solver	
	Pressure-velocity coupling (CFL = 10 and URF = 1.0)	
	2nd order discretization schemes	
	Gradients with least square cell based	

Continuity, energy and momentum equations for the three Cartesian components of the velocity have been computed. Numerical residuals of 10^{-6} for continuity and momentum equations and residuals of 10^{-9} for energy equation have been considered as a condition of convergence but not self-sufficient. Local velocities and

Table 3.3 Boundary conditions used in the 9 CFD simulations used for comparison with experiments

Sim. ID	Inlet velocity [m.s^{-1}]	Fluid inlet Temp. [°C]	Ambient Temp. [°C]	Turns heat source [kW.m^{-3}]
EXP1	1.0 ± 0.4	63.2	21.0	842.05
EXP2	1.0	64.1	20.9	592.37
EXP3	1.1	44.0	20.5	399.06
EXP4	1.5	61.8	17.6	826.43
EXP5	1.4	51.9	17.4	593.03
EXP6	1.6	43.6	17.4	400.32
EXP7	1.8	64.8	18.8	823.57
EXP8	1.8	55.7	19.0	606.19
EXP9	1.8	45.7	19.1	404.66

temperatures in the bottom and top oil pools have been also monitored and observed to be varying, between iterations, on the same order of magnitude as the residuals.

The boundary conditions used in these CFD simulation are listed in Table 3.3.

Table 3.3 indicates the oil velocity magnitude imposed in the two bottom inlets in each simulation. Due to the oscillatory flow meter signal exhibited in EXP1 (Fig. 3.8) this simulation has always been conducted with three different flow rate values. The fluid temperature imposed at the inlet corresponds to T_1 as measured in steady-state and the ambient temperature corresponds to T_5 as measured in steady-state. The ambient temperature is included in the boundary condition of the external walls in contact with air wherein an average heat transfer coefficient has been set. The different values of the ambient temperature slightly affect the heat flux predicted to be transferred to the surrounding air.

The heat generation is uniformly imposed in each of the 48 turns and corresponds to the $\overline{P_{coil}}$ as measured after dividing by the copper volume of the coil.

The relevant physical properties of the cooling fluid considered in these simulations are summarized in Table 3.4.

The dynamic viscosity, density, specific heat and thermal conductivity expressions have been implemented using interpreted User Defined Functions. All the expressions have been truncated for temperature values below 273.15 K and above 473.15 K.

The commercial reference of the cooling fluid used is the same as in the other CFD models reported along this thesis (Nynas Taurus®). This corresponds to the latest data available from the supplier concerning the stock used in the experiments and all the properties have been considered temperature dependent:

- The dynamic viscosity expression has been obtained through fitting of 8 supplier points at temperatures between 20 and 100 °C. The fitting expression used has a maximum relative error of 0.3% and a coefficient of determination = 1.

Table 3.4 Physical properties of the cooling fluid as implemented in CFD

Type of fluid (Commercial Reference)	Properties
Naphtenic mineral oil (Nynas Taurus®)	Dynamic viscosity, μ [kg.m^{-1}.s^{-1}]
	$\mathrm{Exp}((-20.04413) + (12078.37) \times (1/T(K))$ $+ (-4122209) \times (1/T(K))^2$ $+ (574840600) \times (1/T(K))^3)$
	Density, ρ [kg.m^{-3}]
	$1065.801 - 0.6585 \times T(K)$
	Specific heat capacity, Cp [J.kg^{-1}.°C^{-1}]
	$844.5521 + 3.483425 \times T(K)$
	Thermal conductivity, k [W.m^{-1}.°C^{-1}]
	$0.1562308 - 0.00007857143 \times T(K)$

- The density expression has been obtained through fitting of 3 supplier points at temperatures between 20 and 100 °C. The fitting expression used has a maximum relative error of 0% and a coefficient of determination = 1.
- The specific heat expression has been obtained through fitting of 2 supplier points at 40 and 100 °C. The fitting expression used has a maximum relative error of 0% and a coefficient of determination = 1.
- The thermal conductivity expression has been obtained through fitting of 3 supplier points at temperatures between 20 and 100 °C. The fitting expression used in CFD has a maximum relative error of 0.2% and a coefficient of determination = 1.

The materials of solid components and the respective thermal conductivities are listed separately in Table 3.5.

As the CFD simulations have been conducted in steady-state the remaining physical properties of the solids are not relevant.

Table 3.5 Materials and respective thermal conductivities as implemented in CFD

	Material	Thermal conductivity [W.m^{-1}.°C^{-1}]
Cooper coil	Copper	388.5
	Kraft paper	0.16
	Pressboard	0.16
Insulation frames	Pressboard	0.16
Pressboard blocks	High-Density pressboard/Wood	0.17
Spacers	Acrylic	0.035
Acrylic plate	Acrylic	0.19
Polystyrene plates	Polystyrene	0.035
Steel structure	Construction steel	16.3

3.1.4 CFD Results

The main CFD results for EXP1 conditions (for the average oil flow rate—inlet velocity of 1.0 m.s^{-1}) are presented here in order to discuss observed temperature and flow patterns. Notwithstanding this, the whole group of 9 simulations is comprehensively compared against measurements in the next section.

Figure 3.5 includes the velocity magnitude map at the middle height of the fluid domain.

The velocity magnitude map shows that although the average fluid velocity is 10.5 cm.s^{-1} the local maximum velocities are close to 100 cm.s^{-1} as the flow distribution is not homogeneous. In this geometry, and due to channels with higher flow areas near the insulation frames, the fluid tends to flow preferentially in those regions.

Figure 3.6 shows the temperature map at the middle height of the fluid domain. Only fluid temperatures are shown.

The temperature map in Fig. 3.6 shows a large region of hot fluid in the outermost top curved region of the domain. This a consequence of the preferential fluid flow in the innermost channels of the top curved region. Also, hotter regions,

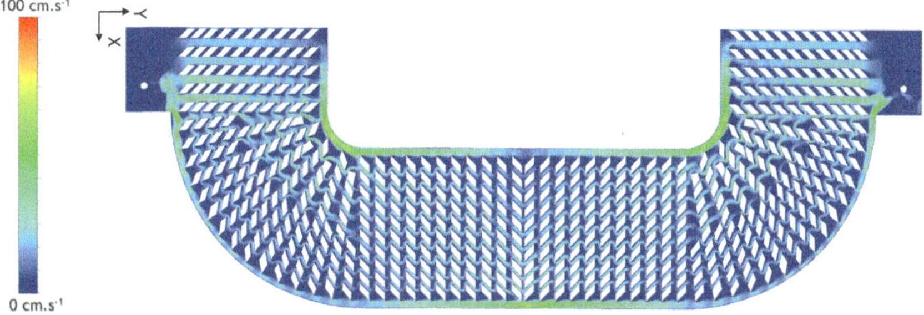

Fig. 3.5 Velocity magnitude map for EXP1 simulation in a plane located at middle height of the fluid channels (Z = −0.001 m)

Fig. 3.6 Temperature map for EXP1 simulation in a plane located at middle height of the fluid channels (Z = −0.001 m)

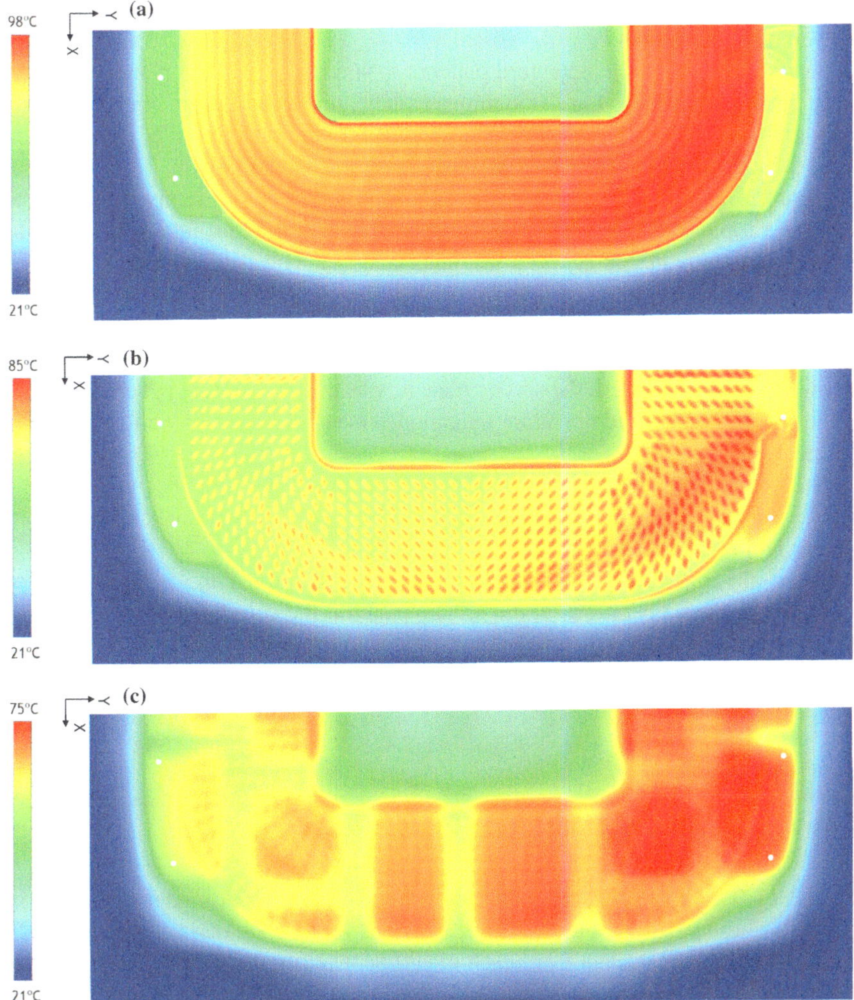

Fig. 3.7 Temperature maps for EXP1 simulation in parallel XY planes: **a** at the symmetry plane (Z = 0.004988 m); **b** at the height of the thermocouples TC1–TC30 (Z = −0.003 m) and **c** at the middle height of the acrylic plate (Z = −0.012 m)

in a smaller extent, are observed near the first and the last row of spacers in the inlet and outlet regions, respectively. This corresponds to regions near a symmetry plane where the oil velocities are quite low.

In this simulation, the heat generated in the portion of the copper being modelled (one quarter of the total) is 737.1 W. A fraction of 53.0 W ($\approx$7.2%) is transferred to the ambient air through the polystyrene plates and the steel structure.

Figure 3.7a shows the temperature map at the *XY* symmetry plane cutting the turns of the copper coil. Figure 3.7b shows the temperature in a *XY* plane cutting the acrylic plate at the Z coordinate where the thermocouples are located. Only solid temperatures are shown.

Figure 3.7a–c show temperature patterns in successive parallel planes. The first plane cuts the copper coil at the symmetry plane, the second plane cuts the acrylic plate at the coordinate where the probe tips of the thermocouples $TC_1 - TC_{30}$ are located and the third plane cuts the acrylic plate at middle height.

Figure 3.7a shows that, despite the observation of hotter oil temperatures in the outermost top curved region, the highest temperatures in the solid are observed to occur in the turns in contact with the inner and outer insulation frames. This is a consequence of the reduced capability of these turns to transfer heat through convection to the cooling fluid.

The comparison between Fig. 3.7a, b shows that, at the coordinate where the thermocouples are located, the acrylic plate retains the temperature pattern observed in the copper coil. The temperature magnitudes measured are lower but the trend is identical.

The complementary analysis of Fig. 3.7c shows that at middle height of the acrylic plate the temperature pattern imposed by the colder external steel structure start to be predominant.

These are important observations which have supported the methodology of measuring local temperatures in the acrylic plate (at the coordinate $Z = -0.003$ m) instead of measuring local temperatures directly in the surface of the copper coil. By employing this methodology, the probability of oil leakages has been greatly diminished.

3.2 CFD Validation

A total of 9 experiments have been conducted in the scale model described in Chap. 2. These 9 experiments have been used to validate the corresponding 9 CFD simulations. The main data from each experiment is summarized in Table 3.6.

Measurements are reported in Table 3.6 under the form—$\bar{x} \pm U_x$. The total combined uncertainties, U_x, have been estimated using the methodology described above. The white shaded columns correspond to the main boundary conditions used in the corresponding CFD models. The grey shaded columns correspond to auxiliary quantities.

In EXP1 the oil flow rate shows the highest total combined uncertainty. The signal from the flowmeter has been observed to have significantly higher oscillations, as shown in Fig. 3.8.

EXP1 results from the combination of the highest power with a low oil flow rate. In this case the Reynolds number in the pipes downstream the oil flowmeter is lower than 1000. This low Reynolds number may not be sufficient to counteract the buoyancy induced movements caused by a stronger heterogeneous temperature distribution inside the pipe. As the flowmeter is measuring the time needed for sound to travel from one transducer to another, and this travel time is sensible to

Table 3.6 Summary of the 9 experiments conducted in the scale model

EXP ID.	T_1 [°C]	T_5 [°C]	Q_{oil} [m³.h⁻¹]	$\overline{u_G}$ [cm.s⁻¹]	P_{coil} [W]	Avg. coil heat flux [W.dm⁻²]
1	63.2 ± 0.6	21.0 ± 0.5	0.42 ± 0.14	10.5 ± 3.6	2948.6 ± 17.7	25.4 ± 0.2
2	54.1 ± 0.5	20.9 ± 0.5	0.41 ± 0.02	10.3 ± 0.6	2074.3 ± 14.0	17.9 ± 0.1
3	44.0 ± 0.5	20.5 ± 0.5	0.46 ± 0.03	11.7 ± 0.9	1397.4 ± 11.0	12.0 ± 0.1
4	61.8 ± 0.5	17.6 ± 0.5	0.61 ± 0.07	15.4 ± 1.7	2893.9 ± 17.3	25.0 ± 0.1
5	51.9 ± 0.5	17.4 ± 0.5	0.59 ± 0.04	14.8 ± 1.1	2076.6 ± 14.0	17.9 ± 0.1
6	43.6 ± 0.5	17.4 ± 0.5	0.65 ± 0.04	16.3 ± 1.0	1401.8 ± 11.0	12.1 ± 0.1
7	64.8 ± 0.5	18.8 ± 0.5	0.74 ± 0.06	18.7 ± 1.4	2883.9 ± 17.3	24.9 ± 0.1
8	55.7 ± 0.5	19.0 ± 0.5	0.75 ± 0.05	19.0 ± 1.3	2122.7 ± 14.1	18.3 ± 0.1
9	45.7 ± 0.5	19.1 ± 0.5	0.76 ± 0.05	19.2 ± 1.3	1417.0 ± 11.0	12.2 ± 0.1

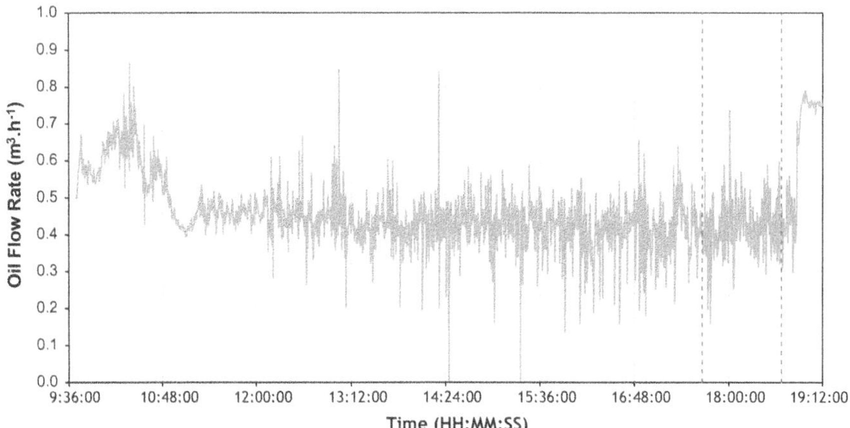

Fig. 3.8 Oil flow rate signal in EXP1

temperature, a noisier signal is collected. Thus, exceptionally for EXP1, 3 CFD simulations have been conducted for the average (Q_{oil}), minimum ($Q_{oil} - U_{qoil}$) and maximum flow rate values ($Q_{oil} + U_{qoil}$) of the whole interval. In the remaining experiments, this oscillation is much lower and a single CFD simulation has been conducted using the average flow rate.

As listed in Table 3.6 the oil temperatures measured in the bottom manifold along with the measured oil flow rate are inputted as boundary conditions in the CFD. The oil temperature at the outlet of the domain is then expected to coincide between the CFD and the experiments. This comparison is plotted for the 9 experiments in Fig. 3.9.

The results shown in Fig. 3.9 have been subdivided in three sets of experiments. Each set corresponds to experiments conducted with identical oil flow rates hence

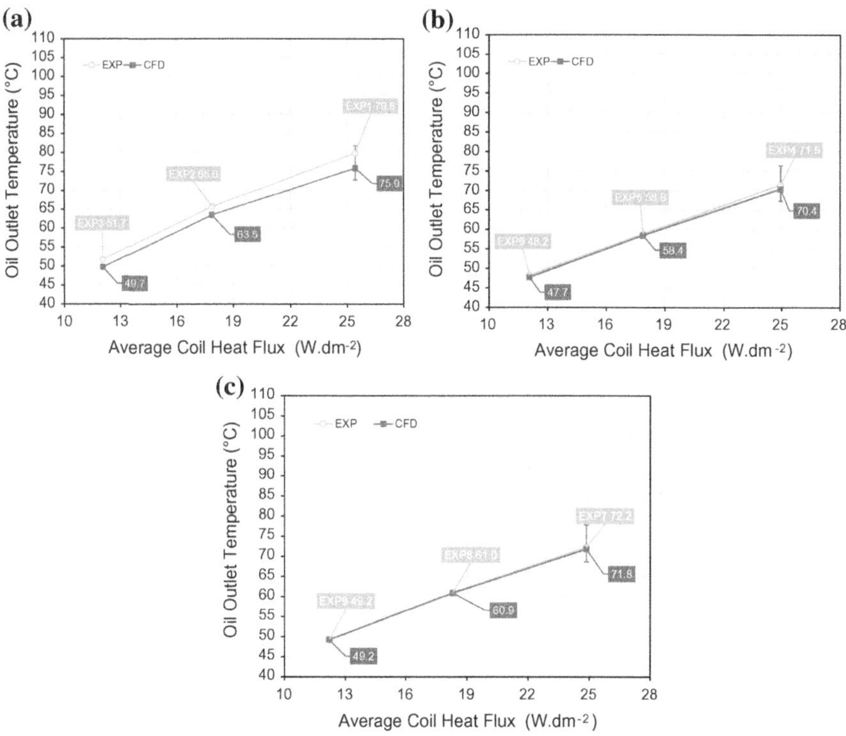

Fig. 3.9 Oil temperature at the outlet. CFD values versus measurements: **a** EXP1–EXP3, **b** EXP4–EXP6 and **c** EXP7–EXP9

with identical average oil velocities. The set comprising EXP1–EXP3 (Fig. 3.9a) correspond to an average oil velocity of 10.9 cm.s^{-1}, the set comprising EXP4–EXP6 (Fig. 3.9b) to an average oil velocity of 15.5 cm.s^{-1} and the set comprising EXP7–EXP9 (Fig. 3.9c) to an average oil velocity of 19.0 cm.s^{-1}. The oil temperatures have been extracted from the CFD simulations using mass weighted integrals evaluated at the outlet surfaces of the domain. From the plots shown it is observable that the oil temperatures predicted using CFD tend to be closer to the experiments for higher average oil velocities. This is probably due to an enhanced oil mixing in the top manifold as the oil velocities increase, which in turn creates an increasingly homogeneous temperature distribution inside the manifold and approximates the measurements from the theoretical CFD values. The results plotted show that the numerical and experimental results are energetically comparable.

Pressure values comparison is plotted, in Fig. 3.10, for the same three sets of experiments.

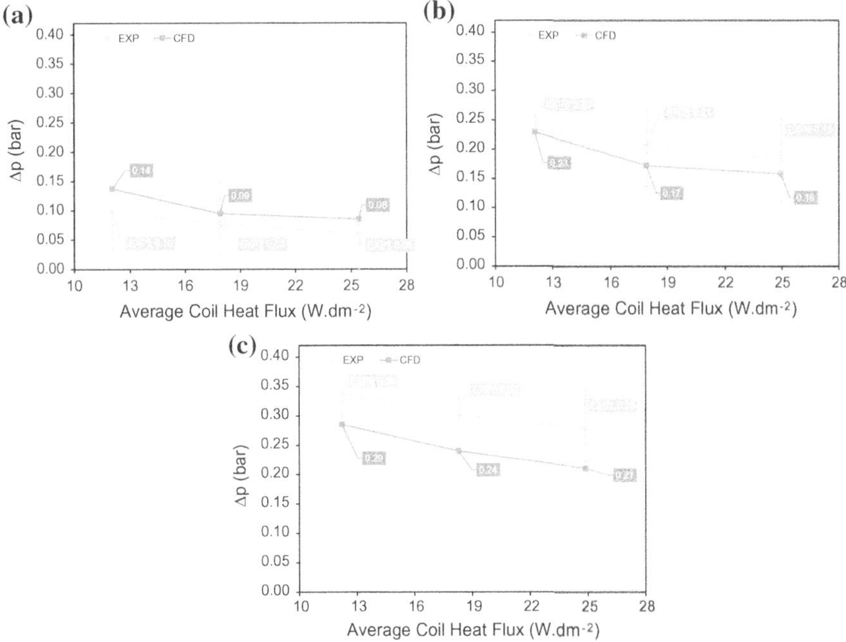

Fig. 3.10 Pressure drop between the bottom manifold and the top manifold. CFD values versus measurements: **a** EXP1-EXP3, **b** EXP4-EXP6 and **c** EXP7-EXP9

The pressure values shown refer to the pressure drop (or difference) between the bottom manifold and the top manifold. Each CFD pressure drop has been computed using

$$\Delta p^{*}_{CFD} = P_{CFD,total,inlet} - P_{CFD,total,outlet} \tag{3.1}$$

where $P_{CFD,total,inlet}$ and $P_{CFD,total,outlet}$ represent the mass weighted integral of the total pressures evaluated at the same inlet and outlet surfaces of the CFD domain where the oil temperatures have also been extracted. These total pressures extracted from CFD concern exclusively static and dynamics effects (the hydrostatic pressures are not considered).

However, the CFD domain does not include the bottom and top manifolds neither the tubes connecting those manifolds to the frontal acrylic plate shown in Fig. 3.11 as red circled regions. An identical arrangement is observed on the top, with shorter tubes between the manifold and the acrylic plates (tubes with 60 mm length in the bottom and 30 mm length in the top). The tubes have an internal diameter of 6 mm and are curved with a long radius of 45°. The 8 globe valves (V4.1 to V4.4 on the bottom and V5.1 to V5.4 on the top) were always simultaneously fully opened.

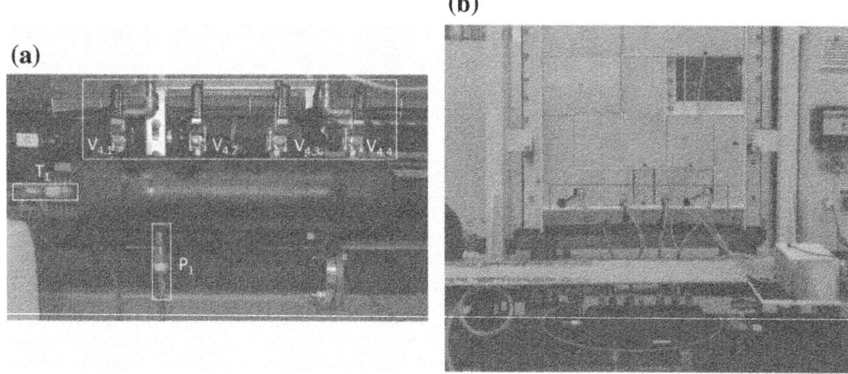

Fig. 3.11 Components not considered in the CFD domain: **a** bottom manifold and **b** tubes

Thus, an additional head loss has been estimated using

$$\Delta p^* = \left[f_b \frac{L_b}{d} + k_{45^\circ} + k_{cont} \right] \frac{1}{2}\rho_1 u_1^2 + \left[f_t \frac{L_t}{d} + k_{45^\circ} + k_{exp} \right] \frac{1}{2}\rho_2 u_2^2 \qquad (3.2)$$

Where u_1 and u_2 are the average oil velocities in each tube respectively bottom and top. The oil flow distribution among the 4 tubes in parallel has been considered uniform.

This equation considers the head loss due to a sudden contraction, k_{cont}, when the oil flow enters each tube, a long radius bend of 45°, k_{45°, and friction in the walls of the tubes, f_b. At the top, besides the friction f_t and a long radius of 45°, k_{45°, a sudden expansion is considered, k_{exp}, when the oil flow from the tubes enters the top manifold. The average Reynolds number in these tubes for EXP1–EXP3 is 1124, for EXP4–EXP6 is 1413 and for EXP7–EXP9 is 1849, hence representing flows in a transition range. Due to this, a simplified Colebrook expression has been used to compute both friction coefficients, f_b and f_t:

$$f_{b/t} = 1.325 \left[\ln\left(\frac{5.74}{Re^{0.9}} \right) \right]^{-2} \qquad (3.3)$$

This expression is known to approximate quite well the Moody Chart values for fully developed turbulent flows in smooth tubes. Finally, the head loss computed using Eq. (3.1) has been added to the total pressure difference computed using Eq. (3.2): $\Delta p_{CFD} = \Delta p^*_{CFD} + \Delta p^*$.

The experimental pressure drop plotted in Fig. 3.10 has been obtained using the following relationship

$$\Delta p_{EXP} = [P_1 - \rho_1 g(h_{max} - h_1)] - [P_2 - \rho_2 g(h_{max} - h_2)] \qquad (3.4)$$

where P_1 and P_2 correspond to the measured relative pressures. These sensors are in the manifolds according to Fig. 2.15. In order to estimate the pressure drop only due to static and dynamics effects, the hydrostatic pressure has been deducted from each measured value.

The pressure comparison plotted in Fig. 3.10, shows a reasonable agreement for EXP1–EXP6. For EXP7–EXP9 the comparison tends to worsen, even though the differences observed are within the total uncertainty range of the measurements. It is noteworthy that the CFD predictions are initially higher than the corresponding measurements and the trend is inverted as the average Reynolds number increases (from 1124 to 1849). This comparison worsens for the last set of experiments comprising EXP7–EXP9. In this case the assumptions of fully developed flow used to estimate the additional head loss may become more relevant. Although, the pressures are not used directly as boundary conditions and the comparison is said to show a reasonable agreement in terms of pressure trends and pressure magnitudes.

Figure 3.12 shows a comparison between the average copper coil temperatures from CFD and average copper coil temperatures obtained from the same three sets of experiments. The CFD values correspond to volume weighted integrals evaluated over all the copper volumes of the coil domain and the experimental values derive indirectly from a measured variation of the ohmic resistance of the copper coil (this latter indirect measurement is described in Chap. 2).

The plots in Fig. 3.12 comprise three series of data: *EXP*, *CFD* and *CFD-Area Corrected*. The series labelled as *CFD* correspond to the current results and the series CFD-Area-Corrected correspond to the same results after applying a correction to account for a higher wetted area than that the area initially designed. Table 3.7 summarizes the average temperatures plotted in Fig. 3.12. The nature of this correction is described below the table.

Accidentally, this copper coil has been manufactured with pressboard strips with 9 mm height (along the Z coordinate) and not 10 mm as initially designed. In Fig. 3.13, the coil is schematically represented as initially designed and as effectively manufactured.

This reduced pressboard strips are shown clearly in Fig. 3.13 using two close photos of the surface of the actual copper coil. As observed in the photos, especially in Fig. 3.13a, the reduced pressboard strips together with the rounded corners of the copper conductors create an additional wetted area which enhances the capability of the coil to transfer heat.

This additional wetted area occurs both in the free regions—without spacers— and in the blocked regions where the spacers are in direct contact with the surface of the turns. The free regions correspond to 68% of the total heat transfer surface of the copper coil and the remaining 32% correspond to the blocked regions. The original CFD model assumes a perfectly smooth copper coil surface, considering each copper conductor with an ideal rectangular cross section and pressboard strips with the same height as the insulated copper bar (along the Z coordinate). Thus,

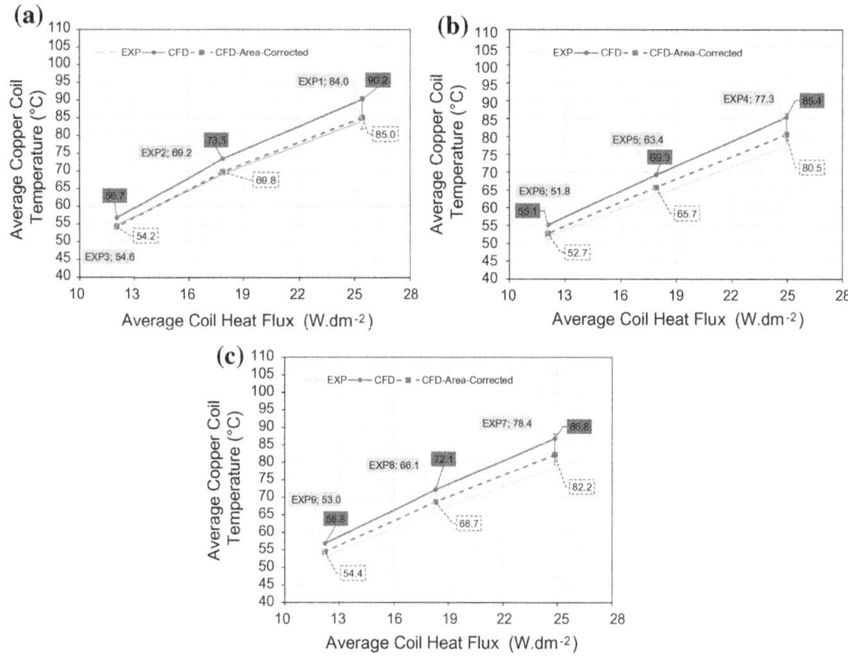

Fig. 3.12 Average Copper Coil Temperatures. CFD values versus measurements: **a** EXP1–EXP3, **b** EXP4–EXP6 and **c** EXP7–EXP9

Table 3.7 Measured average temperatures compared against the CFD predictions

ID	Average temperature, $T_{avg,coil}$		Calculated average temperature, $T_{avg,coil,CFD}$ $(T_{avg,coil,CFD} - T_{avg,coil})$			
	Measured		CFD		CFD-Area-Corrected	
EXP1	84.0 ± 1.9		90.2 (+6.2)	94.1 (+10.1)	85.0 (+1.0)	87.9 (+3.9)
				87.9 (+3.9)		83.2 (−0.8)
EXP2	69.2 ± 2.1		73.5 (+4.3)		69.8 (+0.6)	
EXP3	54.6 ± 2.3		56.7 (+2.1)		54.2 (−0.4)	
EXP4	77.3 ± 1.9		85.4 (+8.1)		80.5 (+3.2)	
EXP5	63.4 ± 2.0		69.3 (+5.8)		65.7 (+2.3)	
EXP6	51.8 ± 2.2		55.1 (+3.3)		52.7 (+0.9)	
EXP7	78.4 ± 1.9		86.8 (+8.5)		82.2 (+3.8)	
EXP8	66.1 ± 2.0		72.1 (+6.0)		68.7 (+2.6)	
EXP9	53.0 ± 2.2		56.8 (+3.7)		54.4 (+1.4)	

pressboard strips with 9.976 mm and not 8.976 mm. The incorporation of this additional wetted area in the 3D CFD model was not considered due to practical constraints, instead the original 3D CFD results have been corrected in order to

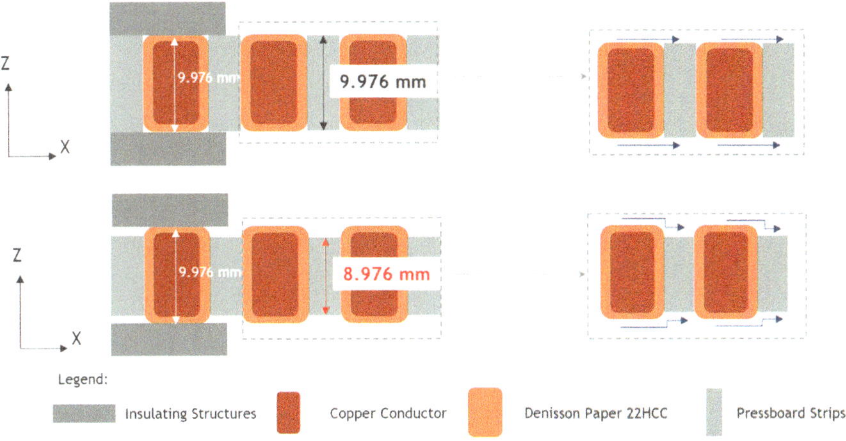

Fig. 3.13 Schematic cut view of the copper coil as initially designed (on the top) and as effectively manufactured (on the bottom)

account for this additional wetted area, finally producing the data series labelled as *CFD-Area-Corrected* in the plots of Fig. 3.12. The correction applied to the original CFD results is further explained.

The relationship between the power of the coil, P_{coil}, and its average temperature, $T_{avg,coil}$, is given approximately by

$$P_{coil} = UA_{designed}\left(T_{avg,coil} - T_{avg,oil}\right) \tag{3.5}$$

where $A_{designed}$ is the wetted area of the copper coil as designed being and $T_{avg,oil}$ is the average oil temperature inside the coil. This is area of the coil modelled in CFD and used to produce the series labelled *CFD*. A similar expression is valid for the actual copper coil

$$P_{coil} = UA_{manufactured}\left(T^{*}_{avg,coil} - T_{avg,oil}\right) \tag{3.6}$$

where $A_{manufactured}$ is the actual wetted area of the copper due to the reduced size of the pressboard strips as depicted in the photos of Fig. 3.14 and $T^{*}_{avg,coil}$ is the corresponding average copper coil temperature of the coil as manufactured.

Assuming identical global heat transfer coefficients in both scenarios, the oil temperatures can also be assumed to be identical, due to the same power being dissipated, thus resulting in the following relationship

$$\frac{A_{designed}}{A_{manufactured}} = \frac{\left(T^{*}_{avg,coil} - T_{avg,oil}\right)}{\left(T_{avg,coil} - T_{avg,oil}\right)} \tag{3.7}$$

Fig. 3.14 Photos of the copper coil surface. EFACEC Courtesy

So the final relationship between designed and manufactured wetted areas for this turn is

$$\frac{A_{designed}}{A_{manufactured}} = \frac{5.976}{(5.976 + 0.5 + 0.5 + 0.32 * 3)} \approx 0.75 \tag{3.8}$$

where 5.976 mm corresponds the length of an individual smooth insulated copper turn as designed: with 3.976 mm plus the additional 2 mm of each pressboard strip. The 0.5 mm correspond to the additional vertical surface (along the Z coordinate) which is wetted due to the reduced size of the strip and 0.32*3 mm corresponds to the additional wetted area opened beneath each spacer (before it did not exist).

Hence, to plot the results shown in Fig. 3.12, the temperature differences between the average copper coil and the average oil temperature have been multiplied by this area correction factor of 0.75. The same temperature offset (assuming linearity) has been applied to the CFD temperatures extracted from the frontal acrylic plate to compare with the local temperature measurements.

After applying this correction, a good agreement between the measured temperatures and the CFD results is observed. The CFD series tend to be systematically above the experiments. Despite this, the final differences observed are within the range of the total combined uncertainties associated with the measurements, hence validating the CFD predictions. Besides this, the trend of the average temperatures for the different heat flux and oil flow conditions can be predicted with confidence.

The plots in Fig. 3.15 show the comparison over the first set of experiments (EXP1–EXP3) for local temperatures. The temperature measurements of the 30 thermocouples previously identified in Fig. 2.12 ($TC_1 - TC_{30}$) are compared against the correspondent CFD predictions ($TC_{CFD,1} - TC_{CFD,30}$).

The plots in Fig. 3.15 comprise again three series of data: *EXP*, *CFD* and *CFD-Area-Corrected*. This latter series correspond to local temperatures in the acrylic that have been corrected with the factor deduced above for the average temperatures. It is also noteworthy the fact that each CFD temperature corresponding to a certain Position ID in the X-axis has uncertainty bars associated. This derives from the challenge of comparing point-to-point temperatures computed numerically with measurements obtained using sensors theoretically located in that positions. Unfortunately (or fortunately) nature is not discrete—it is a continuum—so it is not possible to isolate a discrete location in the X,Y,Z space and in that way isolate the measurement from any other influence.

In order to overcome this challenge, a sample of 15 temperatures has been extracted instead of one single temperature value. These 15 temperatures include the theoretical location but also other 14 temperatures near the theoretical location. As shown in Fig. 2.12 the cylindrical blind holes drilled in the frontal acrylic plate have a variable depth of 17 or 19 mm and a diameter of 6 mm. Hence it is expected that the temperature being measured by each thermocouple is a consequence of several temperatures around the probe tip. Figure 3.16 shows one of those blind

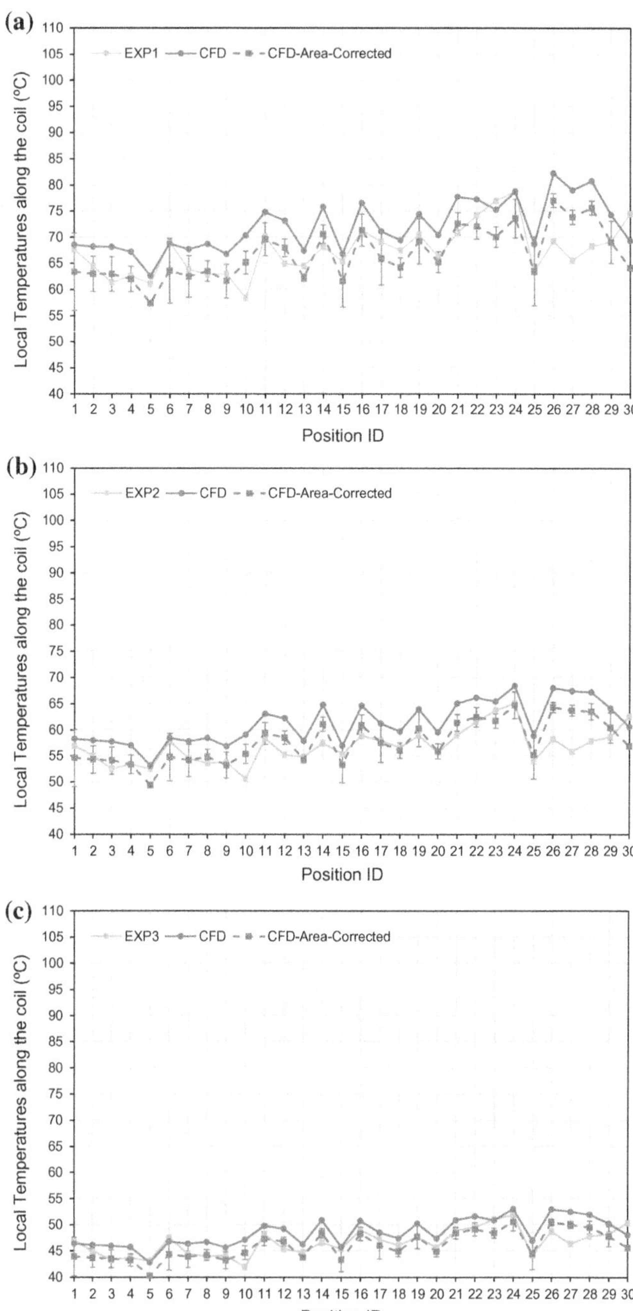

Fig. 3.15 Local acrylic temperatures. CFD values versus measurements: **a** EXP1, **b** EXP2 and **c** EXP3

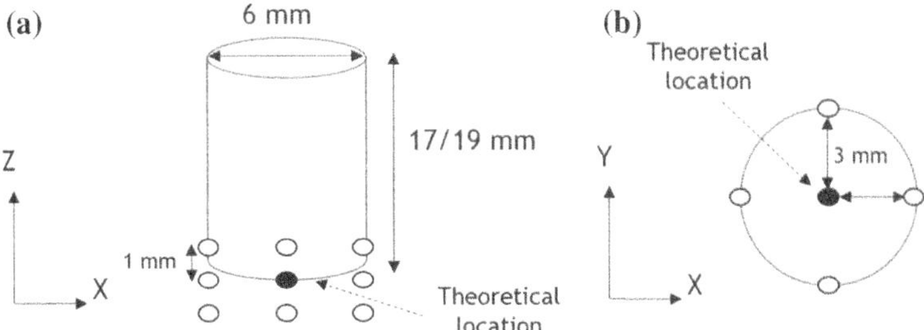

Fig. 3.16 Schematic representation of the blind holes indicating the locations from where temperatures have been extracted in each CFD simulation: **a** lateral view and **b** top view

holes indicating the location of the 15 points wherein local temperatures have been extracted from each CFD simulation.

Figure 3.16a includes a view along the depth of the blind hole and Fig. 3.16b includes a view from the top. As seen in Fig. 3.16a three layers of 5 points each have been considered. These layers are located within ± 1 mm range. Figure 3.16b shows how these 5 points of each layer are equidistant from each other along a circle with 6 mm diameter which corresponds to the physical dimension of the hole. The black filled circles correspond to the theoretical location of the probe tip, while the remaining white filled circles correspond to the 14 X, Y, Z locations from where temperatures have been extracted.

Each blind hole has a single thermocouple inside. Thus, Table 3.8 summarizes the minimum, average and maximum CFD temperatures for each blind hole and for EXP1 resulting from the extraction of 15 temperatures for each hole as in Fig. 3.16.

According to the distribution of thermocouples depicted in Fig. 2.11 the first 15 sensors are located in the bottom part of experimental setup and the last 15 sensors are located symmetrically in the top part. Table 3.9 has been split in two halves according to this distribution: $TC_{CFD,1-15}$ are shown in the left columns and $TC_{CFD,16-30}$ are shown in the right columns. The maximum, minimum and average temperatures are shown for each location based in the sample of 15 points. The respective standard deviation, σ, complements the analysis for each location. The green shaded locations in Table 3.8 refer to locations under inner or outer insulation frames where bigger gradients are expected and where bigger dispersions are observed. Standard deviations tend to be higher from $TC_{CFD,16-30}$ while comparing with $TC_{CFD,1-15}$.

The comparison between CFD predictions and local acrylic temperatures shown in Fig. 3.15 shows a good agreement in terms of trend and magnitudes. The positions where the higher deviations are observed correspond to the last 5 thermocouples located on the top near the exit of the copper coil ($TC_{CFD,26-30}$). These are the regions where the highest temperatures occur and where CFD assumes a perfect condition of symmetry. Due to the symmetry condition, in CFD, these areas tend to shown quite low oil flow rates which artificially induces higher local temperatures than those really measured.

Table 3.8 Summary of the local temperatures extracted from the CFD simulation of EXP1 (for the maximum oil flow rate—$Q_{oil} + U_{qoil}$)

$TC_{CFD,x}$	Temperatures [°C]				$TC_{CFD,x}$	Temperatures [°C]			
	Max.	Min.	Avg.	σ		Max.	Min.	Avg.	σ
1	75.9	64.4	68.5	3.2	16	79.5	70.0	76.5	3.5
2	71.5	65.5	68.1	2.0	17	76.1	67.4	71.1	3.0
3	71.4	65.9	68.1	1.7	18	71.2	68.4	69.4	0.8
4	69.6	65.7	67.2	1.0	19	78.7	70.5	74.3	3.3
5	62.8	62.3	62.5	0.2	20	72.4	66.5	70.4	2.3
6	74.9	64.0	68.7	3.8	21	79.8	75.0	77.7	1.9
7	71.6	64.4	67.6	2.3	22	79.7	73.8	77.2	2.0
8	70.6	67.2	68.7	1.0	23	77.2	74.2	75.2	0.9
9	70.0	64.4	66.7	1.9	24	82.5	73.7	78.8	2.9
10	72.5	67.5	70.3	1.9	25	75.2	65.2	68.7	3.5
11	77.9	68.1	74.7	3.5	26	83.5	80.0	82.2	1.2
12	74.8	71.1	73.1	1.1	27	80.4	77.5	79.0	1.0
13	68.1	66.8	67.4	0.4	28	82.1	78.0	80.7	1.3
14	77.5	73.4	75.7	1.3	29	78.2	72.0	74.2	2.0
15	71.7	64.2	66.8	2.4	30	69.8	69.0	69.3	0.2

Table 3.9 Positions over the frontal acrylic plate where the CFD predictions deviate less than 3°C and more than 3°C. List of the locations with the highest deviations

EXP ID	Criteria		Position IDs where the criteria are not observed
	$|\Delta| \leq 3°C$	$|\Delta| > 3°C$	
1	24 (80%)	6 (20%)	10, 23, 26, 27, 28, 30
2	26 (87%)	4 (13%)	26, 27, 28, 30
3	28 (93%)	2 (7%)	10, 27
4	23 (77%)	7 (23%)	10, 14, 21, 26, 27, 28, 30
5	27 (90%)	3 (10%)	26, 27, 28
6	30 (100%)	0 (0%)	NA
7	23 (77%)	7 (23%)	10, 12, 14, 21, 26, 27, 28
8	27 (90%)	3 (10%)	26, 27, 28
9	30 (100%)	0 (0%)	NA

In fact, during EXP1 and since the refractive index is temperature dependent, certain buoyant plumes of hot oil have been clearly observed in this region (through the acrylic plates). These buoyant plumes tend to disappear either as the inertia of the oil is increased for higher oil flow rates or whenever the average heat flux in the copper coil is diminished. These buoyant driven streams of oil exhibit clear random movements, hence overruling the mathematical CFD assumption of symmetry. This reasoning seems to explain why the temperature differences in $TC_{CFD,26-30}$ tend to be systematically higher in CFD than in the experiments.

To have a broader statistical perspective over the CFD accuracy, a criterion of $\pm 3\ °C$ has been considered sufficient to qualify the measurement as accurate. Table 3.9 summarizes the number of positions wherein this criterion is observed for the all experiments conducted from EXP1 to EXP9.

Over the clear majority of the positions along the acrylic plates, the CFD predictions show an absolute difference against the measurements below or equal 3 °C. For EXP1, EXP4 and EXP7 this criterion is observed in 80, 77 and 77% of the positions. And these experiments correspond to the 'worst' operating conditions where stronger buoyancy is expected due to higher average heat fluxes in the copper coil. For the remaining EXP2, EXP3, EXP5, EXP6, EXP8 and EXP9 this criterion is observed in 87, 93, 90, 100, 90 and 100% of the positions. As above mentioned, the positions where the criterion is not observed are systematically located between 26 and 30, $TC_{CFD,26-30}$.

3.3 Conclusions

The CFD is considered the most accurate numerical approach to model the thermal performance of shell-type coil/washer systems. In Chap. 4 CFD is used to extract correlations that were further included in the FluSHELL tool (correlations for the friction and heat transfer coefficients) and in Chap. 5 the performance of the FluSHELL tool is compared against sets of CFD simulations. Consequently, it is of upmost importance to validate the CFD results.

For this purpose, a detailed CFD model has been described. This CFD model intends to represent the experimental setup described in Chap. 2. A group of 9 simulations using this model were conducted and compared with the respective 9 experiments.

After a comprehensive comparison:

- The CFD has been able to predict the average temperatures with an absolute average deviation of 1.8 °C and an absolute maximum deviation of 3.8 °C in EXP7. The average deviation of 1.8 °C is within the range of combined uncertainty of the measurement methodology which is $\pm 2.1°C$.
- As described in Chap. 2, 30 thermocouples have been inserted in blind holes drilled in the frontal acrylic plate. For 26 of them, the CFD predictions show deviations lower than 3 °C. Moreover, the local temperature trends and magnitudes predicted by CFD show good agreement with measurements and reinforce the previous observation.
- The CFD results seem energetically consistent. Based on the inlet temperature condition which was imposed in every simulation, the CFD is expected to compute a higher outlet oil temperature that results from the balance of the heat transferred from the copper coil to the oil and thus transferred from the oil to the

ambient air. As the coil/washer system is not perfectly adiabatic this balance entails an additional uncertainty. Nevertheless, the CFD computed oil temperatures at the outlet is highly coherent with the oil temperature measured in the top manifold. The CFD predicted oil temperatures at the outlet show an absolute deviation of 1.2 °C.

Consequently, the CFD results reported are claimed to be adequately validated from an average temperature perspective and from a local temperature perspective.

Furthermore, it has been observed that an accidental reduced size of the pressboard strips (of 1 mm) creates an additional cooling mechanism that lowered significantly the measured temperatures. Although by accident, this can be itself a relevant finding to underpin further research about more efficient geometric arrangements of the copper coil.

Chapter 4
The FluSHELL Tool

FluSHELL is a thermal-hydraulic network tool capable of modelling and simulating the steady-state temperature and velocity distribution of the cooling fluid inside the coils of shell-type electrical transformers. In this chapter the FluSHELL tool is described.

In Sect. 4.1 the FluSHELL tool is conceptually described along with the logical identification of the models that together form the FluSHELL tool. In Sect. 4.2 those models are individually addressed comprehensively. In this section the organization of the algorithm is depicted and the main numerical methods used to solve the resulting set of equations are identified.

In Sect. 4.3 the empirical correlations used in FluSHELL to compute transport properties, such as the friction factors and heat transfer coefficients, are obtained using detailed 3D CFD simulations.

The main outputs of the tool are shown in Sect. 4.4 while Sect. 4.5 summarizes the main characteristics of the tool and outlines conclusions about its development.

4.1 Introduction

An electrical transformer is a complex system where the main component is a copper coil through which electrical current flows. Thus, an electromagnetic field is formed around it which causes additional induced electrical currents and mechanical forces. In turn, the flow of electrical currents through the metallic conductors of the coil promote the generation of heat that is then transferred to a cooling fluid in order to maintain the temperatures of the coil below certain steady and acceptable magnitudes (Blume et al. 1951; Del Vecchio et al. 2001).

FluSHELL is an engineering software tool which aims to describe accurately the spatial distribution of the temperatures both in the coils and in the surrounding cooling fluid inside the typical coils of large shell-type power transformers. For this purpose, FluSHELL computes iteratively the spatial distribution of pressures and

© Springer International Publishing AG 2018

H. Campelo, *FluSHELL – A Tool for Thermal Modelling and Simulation of Windings for Large Shell-Type Power Transformers*, Springer Theses, https://doi.org/10.1007/978-3-319-72703-5_4

velocities in the internal cooling fluid as well as the corresponding temperatures within the solid components of interest. As early reported by Montsinger in the late 1920s (Montsinger 1930) and still used in the most recent worldwide standards (IEC 2005), by every 6 °C temperature increase, the lifetime of the equipment can be halved. As transformers represent large capital investments, the economical relevance of Montsinger observation and the subsequent benefits associated with increased capabilities of predicting temperature more accurately have been underpinning several research activities on this subject (Picher et al. 2010; Tanguy et al. 2004).

Earlier thermal modelling approaches focused on developing simpler formulas with adjustable parameters which could be confidently correlated with measurements, especially over decades of practical experience (Del Vecchio et al. 2001). This approach can be acceptable for standard designs but less interesting for new or untried transformer designs. Over the 1980s and 1990s, together with faster and affordable computing capabilities, thermal-hydraulic network algorithms emerged (Declercq and van der Veken 1999; Oliver 1980; Radakovic and Sorgic 2010) enabling predictions with more detail and minimizing progressively the number of adjustable parameters. Over the last decade Computational Fluid Dynamics (CFD) is becoming more common and its application to power transformers is driving the comprehension of the underlying cooling mechanisms to unprecedented levels (Campelo et al. 2013; Kranenborg et al. 2008; Schmidt et al. 2013; Skillen et al. 2012; Torriano et al. 2010; Weinlader and Tenbohlen 2009; Yatsevsky 2014). Currently most industry experts recognize that, due its mathematical and numerical robustness, CFD is the most powerful technique available. CFD simulations usually involve a much finer spatial discretization and enable the computation of the governing differential equations without simplifications. However, its direct integration in the design-cycles is not envisaged in the mid-term (Cigre 2016) and CFD is currently regarded as a virtual R&D laboratory used to extract relevant information to validate or to include in simpler modelling approaches (Campelo et al. 2012; Coddé et al. 2015; Wu. et al. 2012a, b).

Due to this, the thermal-hydraulic network algorithms are believed to encompass a reasonable compromise between accuracy, effort and perhaps independence. And, in this context, the FluSHELL tool embodies the first known thermal-hydraulic network algorithm for shell-type transformers.

In this work, the thermal-hydraulic sub-system of coil/washer system under study is decoupled from its correspondent electromagnetical and mechanical sub-systems:

- On one hand, once the heat generated in a coil is known, the backward dependency between the thermal-hydraulic sub-system and the electromagnetic field is not considered;
- On the other hand, once the geometry of the coil is designed to withstand certain mechanical stresses, the interdependency between the thermal-hydraulic sub-system and the mechanical forces is also not considered.

If needed, and relevant, all these combined sub-systems might be addressed at a wider system level by coupling the FluSHELL tool with other numerical tools dedicated to each purpose. Providing means for this has been set as one the specific objectives, however this integration is beyond the scope of the present work.

To decouple the thermal-hydraulic sub-system from the electromagnetical and mechanical sub-systems, the coil/washer system is modelled and simulated by itself imposing suitable boundary conditions either for pressure or mass flow rate at the inlets and imposing the total amount of heat generated in the coil.

4.2 FluSHELL Description

The FluSHELL tool describes the thermal-hydraulic behavior of a single coil/washer system coil using a set of algebraic equations which are analogous between several sub-systems. Although, the FluSHELL tool can be applied to any coil/washer system, once its geometry is known. Here the tool is described using the geometry of the coil/washer system from the experimental setup built and described in Chap. 2.

The equations, in these analogous sub-systems, can be transformed into their equivalents by changing the variables that express quantities, as shown in Table 4.1.

Table 4.1 emphasizes that Newton's Law and Darcy's Law share the same structure of Ohm's Law. Therefore, the heat flow rate, Q, and the mass flow rate, q, are directly proportional to the potential of each sub-system, expressed in terms of temperature and pressure (respectively) while their relation assumes the form of a conductance.

FluSHELL tool is based on applying this analogy to the methodologies used for electrical circuits, particularly with electrical circuits, where the resistances and voltages for each branch of that circuits are determined by the application of Kirchoff's circuit laws. To put this concept in practice, the primary major step involves the discretization of the coil/washer system into sets of nodes interconnected by branches. First the coil/washer system is decomposed into fluid and solid domains.

In the fluid domain, the branches correspond to channels through which the fluid flows and the nodes correspond to the intersections (or junctions) of different

Table 4.1 Thermal-hydraulic-electrical analogy

Quantity	Thermal sub-system	Hydraulic sub-system	Electrical sub-system
Potential	ΔT [°C]	ΔP [Pa]	ΔV [V]
Flow	Q [W]	q [kg.s^{-1}]	I [A]
Resistance	R^t [W^{-1}.°C]	R^h [Pa.kg^{-1}.s]	R^v [V.A^{-1}]
Conductance	$C^t = \frac{1}{R^t}$ [W.°C^{-1}]	$C^h = \frac{1}{R^h}$ [Pa^{-1}.kg.s^{-1}]	$C^v = \frac{1}{R^v}$ [V^{-1}.A]

channels. Figure 4.1a shows the fluid domain simulated in FluSHELL, Fig. 4.1b shows the partition of the cooling fluid flow area into different channels and Fig. 4.1c depicts the respective branches and nodes where the cooling fluid circulates.

In Fig. 4.1c the nodes are represented by the filled markers and the branches are represented by the line segments.

In the solid domain, the nodes correspond to certain locations (points) in each coil and the branches correspond to possible paths through which the heat generated can be transferred (e.g. through thermal conduction and thermal convection- the thermal radiation mechanism in the coil/washer system is negligible).

Figure 4.2 depicts the branches, in the main solid domain of the coil/washer system, comprising 48 turns where the heat is generated. A solid domain representing the coil of a commercial shell-type transformer would comprise 20–150 turns. In Fig. 4.2 each single turn is represented by consecutive line segments which correspond to partitions of each single turn. These partitions occur when the line representing the turn intersects either a fluid channel, a spacer or an insulation frame. These line segments corresponding to the partitions of each turn are the unitary solid domains considered in FluSHELL. Analyzing an individual turn, the center of each line segment represents a node and the connection between two nodes can be similarly understood to be a branch.

FluSHELL discretizes the coil/washer system into a group of two coupled networks: the fluid network and the solid network. These two networks are said to be

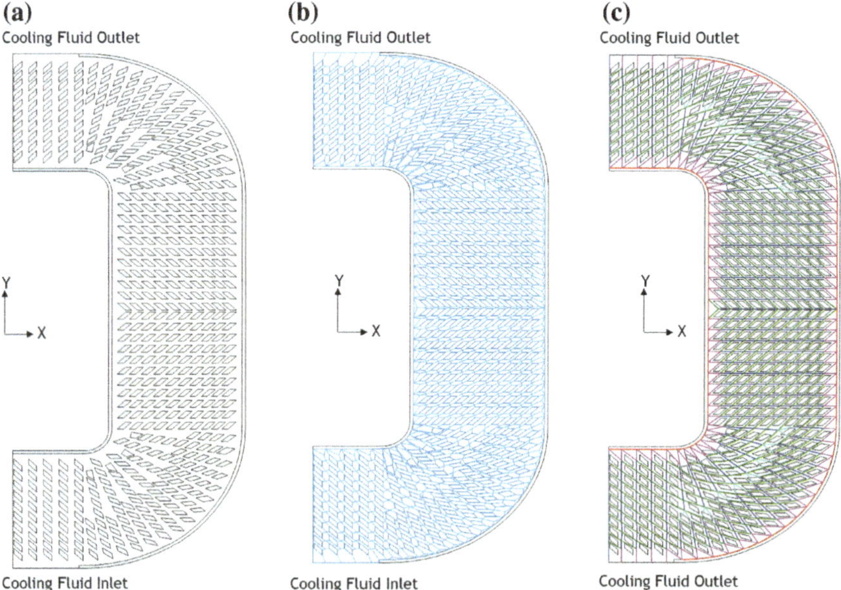

Fig. 4.1 FluSHELL fluid domain: **a** washer with spacers and with the insulation frames; partition into channels; **c** nodes and branches

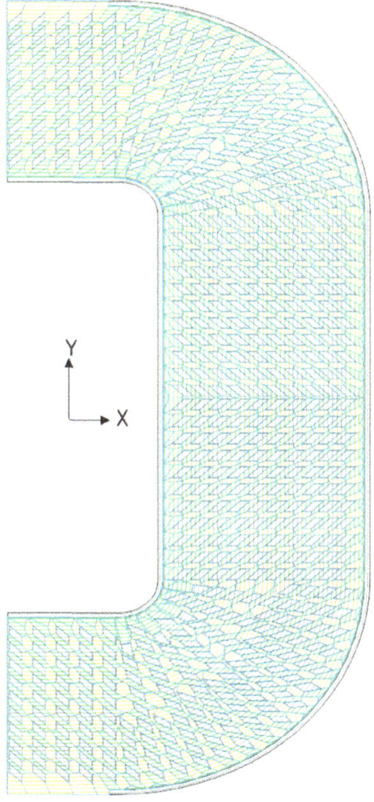

Fig. 4.2 Branches of the solid domain represented overlapping the fluid channels, the spacers and the insulation frames

coupled because the nodes of the two networks are not directly connected, instead the networks communicate and interchange data. Therefore, the networks cannot be solved separately and the overall solution in FluSHELL is obtained through an iterative computation.

The sequential diagram in Fig. 4.3, describes the main models and logical blocks involved in the FluSHELL tool (from the upstream geometric input data towards the downstream spatial distribution of temperatures and fluid velocities). Each model identified in this diagram is discussed in different subsections of the Sect. 4.2.

As depicted in Fig. 4.3, FluSHELL depends on the upstream data that describes the geometry of the coil/washer system. Based on this geometry, a topological model generates both the fluid and the solid networks. The generation of these networks and its configurations are described in Sect. 4.2.2.

Both networks are then combined with the boundary conditions and an initial guess of the temperatures, before being both loaded into an iterative loop of two additional models:

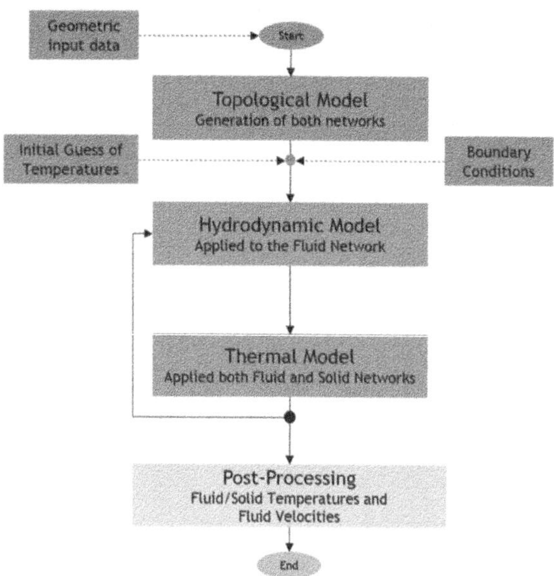

Fig. 4.3 Sequential diagram of FluSHELL modelling steps

1. The **Hydrodynamic Model** primarily interprets the fluid network, determines the hydraulic conductance associated with each fluid channel and computes the mass flow rate associated with each branch along with the corresponding pressures in the intersections between channels (nodes). The Hydrodynamic Model is described in Sect. 4.2.3.
2. The **Thermal Model** further interprets the solid network, determines the thermal conductance associated with each branch and computes its temperatures. Some of the branches of the solid network depend on the fluid temperatures and fluid velocities in the adjacent fluid channels, meaning that the loop between the two models is repeated until predefined criteria of convergence are observed. The Thermal Model is described in Sect. 4.2.4.

Finally, a considerable amount of data is generated corresponding to the temperatures and mass flow rates in each branch of the fluid or solid networks. A post-processing algorithm manipulates the data from each branch of both networks and reports the relevant information, either visual or numerical. The main outputs of the FluSHELL tool are identified and described in Sect. 4.4.

4.2.1 General Description

As depicted in Fig. 4.3, the FluSHELL tool comprises three models:

- the Topological Model where the networks needed are generated;
- the Hydrodynamic Model where the pressures and mass flow rates are computed;
- the Heat Transfer Model where the temperatures in the solid and in the fluid are computed according to the mass flow rates computed in the Hydrodynamic Model.

The FluSHELL tool relies upon basic conservation principles such as conservation of energy and conservation of mass. Unlike CFD, wherein the same conservation principles are described by a system of partial differential equations, FluSHELL describes them by simpler algebraic equation systems, and thus time-to-solution is found to be significantly shorter (see Chap. 5). In the next subsections, each of these three models is addressed in detail.

4.2.2 Topological Model

The thermal-hydraulic sub-system of a coil/washer system is composed by a coil and the adjacent fluid domain that is created both by the spacers and by the insulation frames (as described in Chap. 1).

The topological model discretizes (or subdivides) the coil and the adjacent fluid domain. A sequential diagram of the topological model generation is shown in Fig. 4.4.

The topological model starts by reading or importing four groups of input data:

- Information on the coils namely the number, configuration, type and dimensions.
- Information on the global dimensions and geometry of the washer, including the position (and number) of the fluid inlets and outlets as shown in Fig. 4.5.

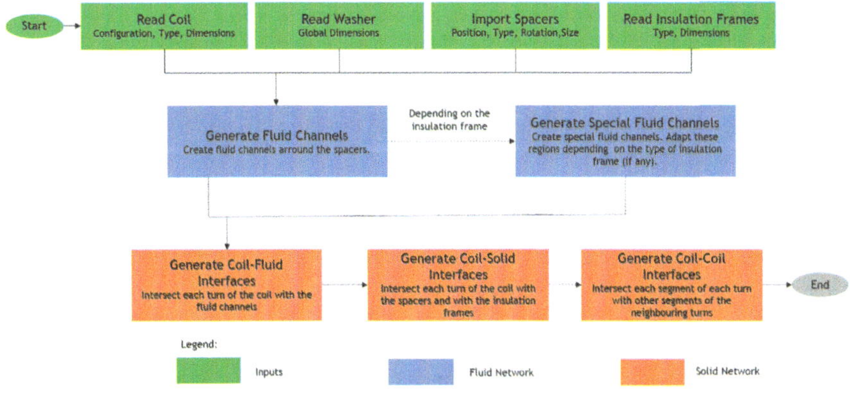

Fig. 4.4 Sequential diagram of FluSHELL topological model steps

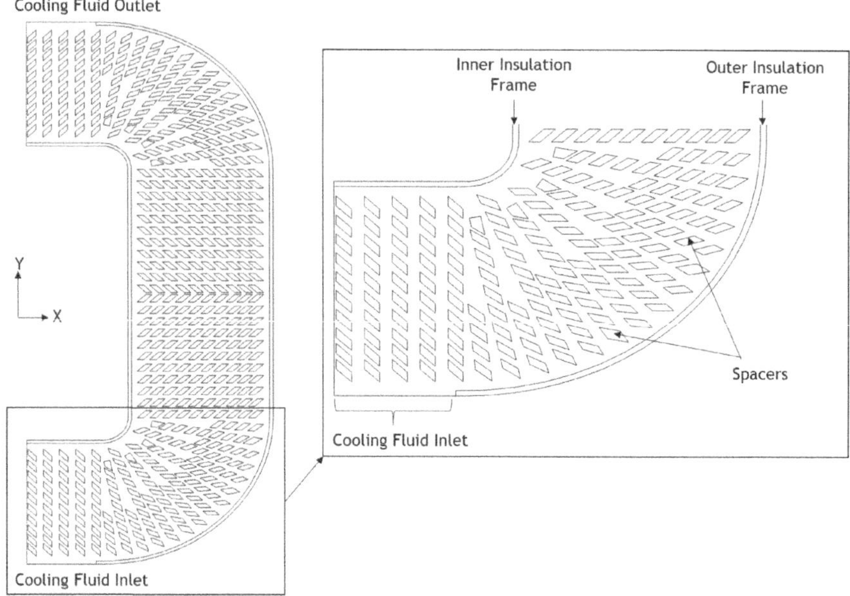

Fig. 4.5 Image of a washer and zoomed view of the spacers and insulation frames confining the fluid flow

- Information on the spacers that support mechanically the windings and simultaneously guide the fluid flow. As shown in Fig. 4.5 these spacers are distributed over the whole washer, generally organized in rows, and there are four shape types common for all transformer designs. Required data for each spacer incudes the centroid coordinates, shape type, rotation and size.
- Information for the inner and outer insulation frames, located at the washer borders as shown in Fig. 4.5 (if any). These frames reduce the fluid flow area in these locations and can be linear shaped, as in the case of Fig. 4.5, or present a dented-like structure as the general case described in Chap. 1. The topological model adapts the fluid network in these locations as a function of the shape of these insulation frames.

Once the geometry is completely defined, the topological model can generate the fluid channels as shown in Fig. 4.6.

Figure 4.6 depicts the fluid channels in an XY plane. The polygons shown are extruded along Z coordinate to generate polyhedral fluid channels. The channels have different shapes or characteristics depending on the location. Channels formed between two consecutive spacers belonging to the same row are called *transverse channels*, while channels formed by spacers between consecutive rows are referred as *radial channels*. The remaining fluid channels are defined as special, because the network needs to be adapted as a function of the shape of the insulation frames or as a function of the number of the fluid inlet channels as shown in Fig. 4.7.

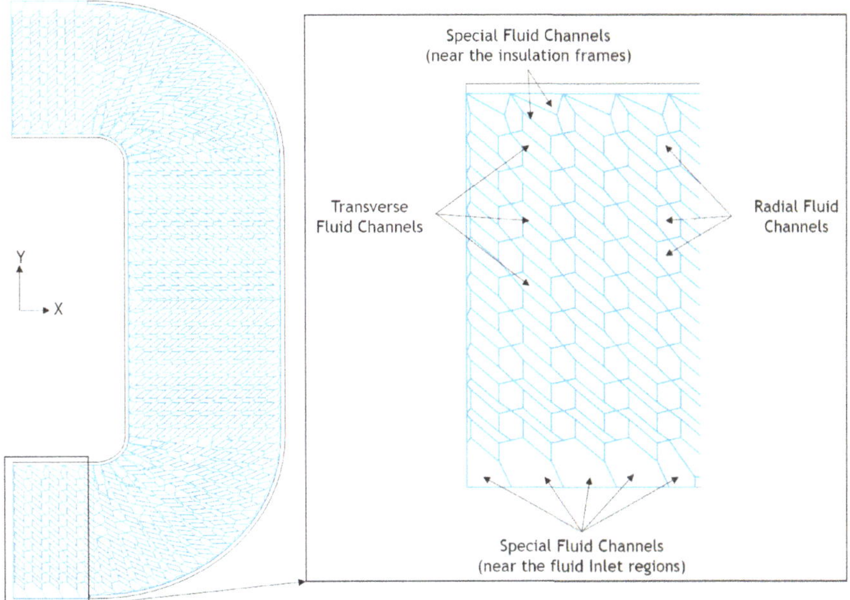

Fig. 4.6 Image of the fluid network generated by FluSHELL

Each fluid channel shown in Figs. 4.6 and 4.7 has five geometry attributes: fluid volume; characteristic length; wetted area; flow area; and hydraulic diameter as listed and defined in Table 4.2.

These geometrical attributes are determined with the topological model, and are latter used for the computation of the hydraulic resistances associated with each fluid channel as described in Sect. 4.2.3.

The FluSHELL tool considers the temperature within each fluid channel to be constant and a single average velocity magnitude is used to characterize the entire channel. This represents a crucial difference to common commercial CFD codes (e.g. ANSYS® Fluent), where each channel is subdivided into several mesh elements resulting in temperature and velocity profiles inside each fluid channel.

Subsequently to the definition of all channels, the topological model numbers each fluid channel and associates it with a branch. In turn, the nodes are obtained as the intersections of the different branches, and are also numbered. The final fluid network comprising the branches, depicted as line segments, interconnected by nodes, represented as full circles, is shown in Fig. 4.8.

Once the fluid network is completely defined, the FluSHELL topological model proceeds to the generation of the solid network that according to the diagram in Fig. 4.4 comprises the generation of **three types of interfaces**:

- **Coil-Fluid interfaces**, represented in Fig. 4.9 by orange and green colored line segments. These correspond to segments of the coil which are in direct contact

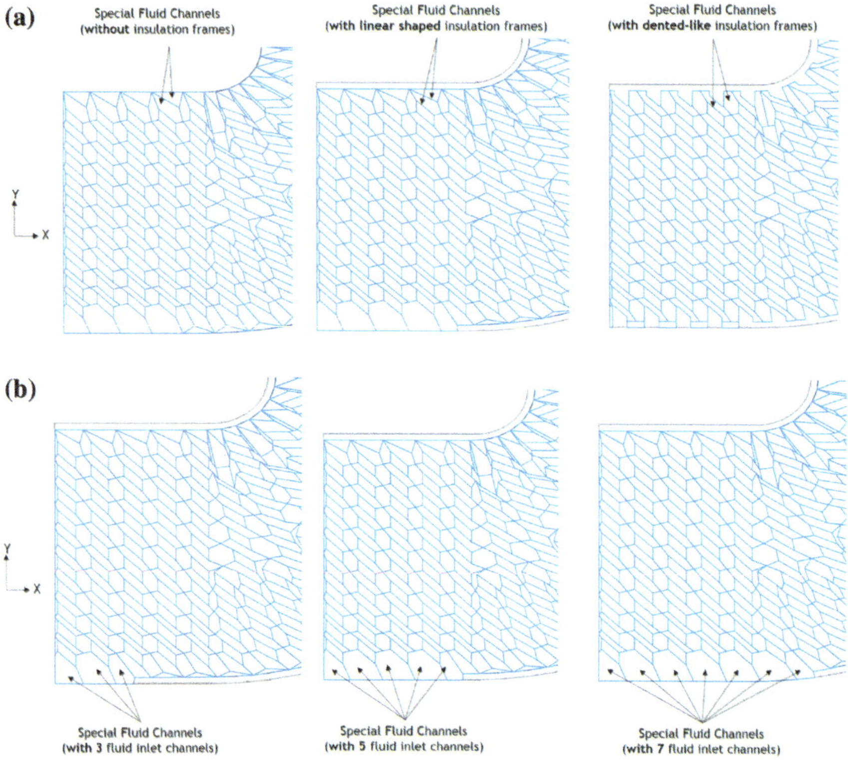

Fig. 4.7 Images of the special fluid channels adapting: **a** to different insulation frames and **b** to different numbers of fluid inlets

with the cooling fluid. Orange colored segments are those in contact with radial fluid channels and the green colored correspond to those in contact with transverse fluid channels.

- **Coil-Solid interfaces**, represented in Fig. 4.9 by grey colored line segments. These correspond to segments of the coil which are in direct contact either with the spacers or with the outer/inner insulation frames.
- **Coil-Coil interfaces**, represented in Fig. 4.10 by black colored line segments. These correspond to segments of a certain turn in direct contact with other segments from other neighbouring turns.

These different types of interfaces enable further implementation of adequate heat transfer resistances for each type of interface. More details are described in Sect. 4.2.4.

Furthermore, the three types of interfaces correspond to the possible paths through which the heat generated is transferred and each path can be understood as a branch. The nodes are located in the middle of each branch. Hence, the branches

Table 4.2 Geometrical attributes of the fluid channels of the fluid network

Attribute	Definition	
Fluid volume, V_{ch}	$V_{ch} = A_{toppolygon} H_{ch}$	(4.1)
	The volume is the area of top six-sided polygon multiplied by the height of the channel	
Characteristic length, L_{ch}	$L_{ch} = \sqrt{(x_4 - x_1)^2 + (y_4 - y_1)^2}$	(4.2)
	The characteristic length is the distance between the vertices 1 and 4 belonging to the top six-sided polygon	
Wetted area, $A_{w,ch}$	$A_{w,ch} = \sum_{i}^{n} A_{wi,ch}$	(4.3)
	The wetted area is the sum of the areas of all the faces of the channel contacting directly with the internal cooling fluid	
Flow area, $A_{f,ch}$	$A_{f,ch} = \frac{V_{ch}}{L_{ch}}$	(4.4)
	The flow area of each fluid channel is the volume of the channel divided by its characteristic length	
Hydraulic diameter, $d_{h,ch}$	$d_{h,ch} = \frac{4 V_{ch}}{A_{w,ch}}$	(4.5)

(and the implicit nodes) identified in Fig. 4.9 and in Fig. 4.10, together form the solid network.

As in the case of the fluid channels, the FluSHELL tool considers the temperature constant within each branch of the solid network. In addition, the heat fluxes from some branches to the cooling fluid are computed using correlations detailed in Sect. 4.3. This represents another crucial difference to common commercial CFD codes (e.g. ANSYS® Fluent), wherein each segment of a turn is subdivided into several mesh' elements resulting in a higher resolution. In CFD codes, the fluid velocity field is computed directly avoiding the use of empirical expressions to compute these heat fluxes.

The topological model just described corresponds to the first layer wherein the thermal-hydraulic system of equations is then built.

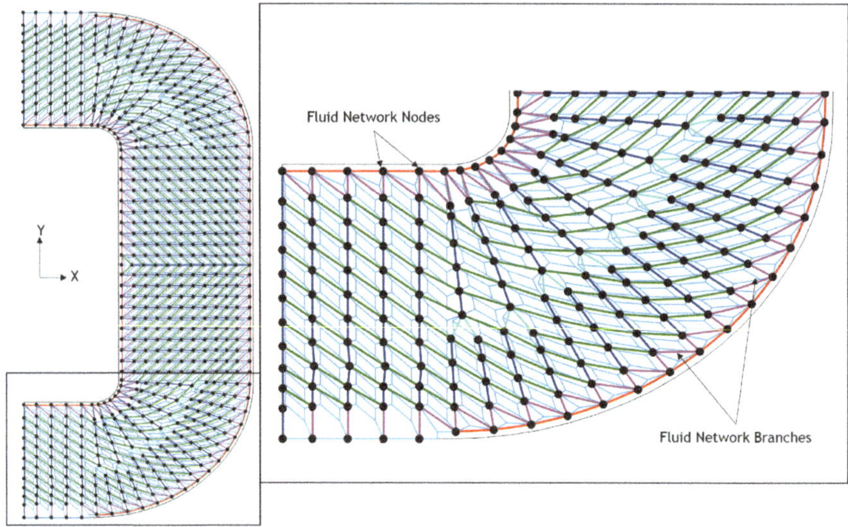

Fig. 4.8 Image of the fluid network of branches and nodes generated by FluSHELL topological model

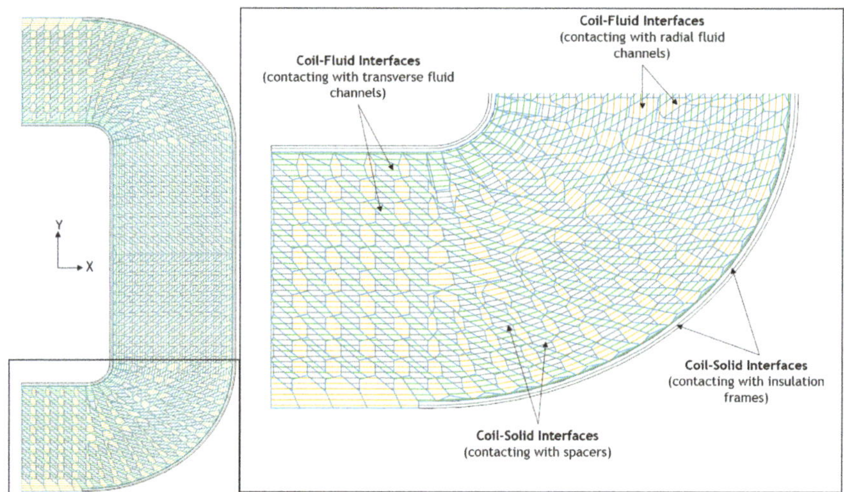

Fig. 4.9 Image of the solid network with coil-fluid and coil-solid interfaces generated by FluSHELL topological model

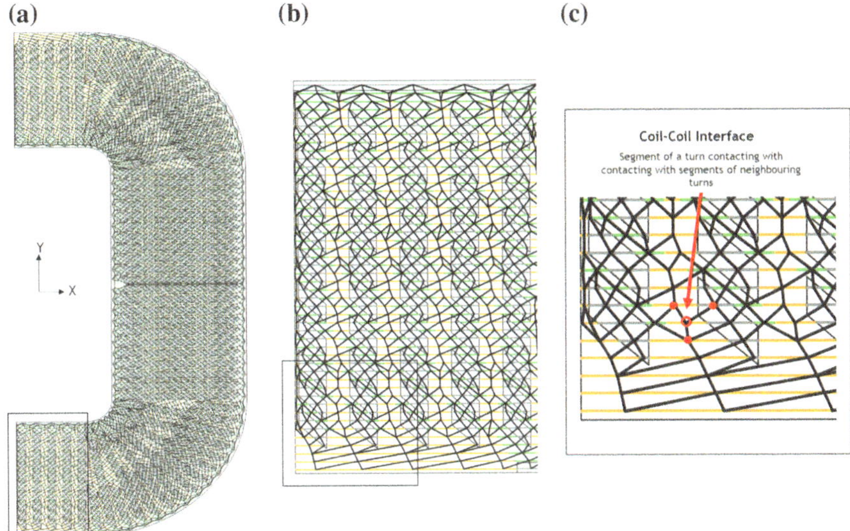

Fig. 4.10 Image of the solid network with coil-coil interfaces generated by FluSHELL topological model. Progressive zoom from **a** to **c**

4.2.3 Hydrodynamic Model

The Hydrodynamic Model comprises the set of equations that describe the relationships between node pressures and branch flowrates over the whole fluid network. The result of this model is the fluid pressure in each node of the network and hence the mass flow rate in each branch. For this purpose, the previous generated fluid network is ordered and numbered, as exemplified in Fig. 4.11 for a region near the inlet nodes.

Fig. 4.11 Fluid nodes and branches numbered (over a region near the inlets)

In Fig. 4.11 the black numbers refer to the nodes, the blue numbers refer to the radial branches and the green numbers refer to the transverse branches.

Therefore, the hydraulic-electrical analogy can be established, as shown in Fig. 4.12 where the pressure, P_k, corresponds to the electrical voltage in node k, and two consecutive nodes, P_k and P_{k+1}, form a fluid channel with a mass flow rate, $q_{k:k+1}$, that corresponds to the electrical current through branch $k : k + 1$.

The model applies Kirchoff's current law to determine the algebraic sum of all branch currents entering or leaving a node. Hydraulically this law means that there is no accumulation of mass in the fluid nodes which guarantees the mass conservation over the whole fluid network.

The mass balance depicted in Fig. 4.12 for node k is given by

$$q_{k-1:k} + q_{k:k+nk} - q_{k:k+1} - q_{k-nk:k} = 0 \tag{4.6}$$

In addition, FluSHELL considers the fluid network purely resistive, hence the pressure difference in a generic branch $k - 1 : k$ is described by

$$\Delta p_{k-1:k} = R^h_{k-1:k} \, q_{k-1:k} \tag{4.7}$$

where $R^h_{k-1:k}$ is the hydraulic resistance of branch $k - 1 : k$. Equation (4.7) can also be written in terms of the hydraulic conductance, $C^h_{k-1:k}$

$$q_{k-1:k} = C^h_{k-1:k} \, \Delta p_{k-1:k} \tag{4.8}$$

If the hydraulic conductance for each branch is known, the application of Eqs. (4.6) and (4.8) to the whole fluid network results in a system of algebraic

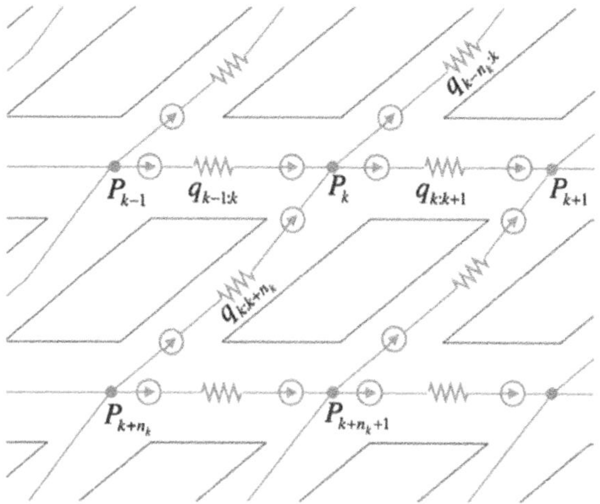

Fig. 4.12 Hydraulic-electrical analogue of the fluid flow around the spacers

equations with the form $\mathbf{Y}\mathbf{p} = \mathbf{q}_s$ where $\mathbf{Y}$ is a banded, symmetric matrix whose coefficients are

- y_{kk} which is the sum of the conductance of all branches connected to node k
- y_{lk} which is the negative of the conductance connecting node l to node k

The vector $\mathbf{p}$ represents the pressures in each node and $\mathbf{qs}$ represents the mass flow sources or sinks at each node (Desoer and Kuh 1969; Dias and Payatakes 1986).

The flow through each channel is assumed to be laminar and the hydraulic resistance for each channel is given by:

$$R^h = \frac{\mu L_{ch} f(Re)\, Re}{2\rho d_{h,ch}^2 A_{f,ch}} \tag{4.9}$$

where ρ is the fluid density, μ is the fluid dynamic viscosity, $d_{h,ch}$ is the channel hydraulic diameter, L_{ch} is the channel characteristic length, $f(Re)$ is the friction coefficient, $A_{f,ch}$ is the channel flow area and Re is the Reynolds defined as

$$Re = \frac{\rho u_{ch} d_{h,ch}}{\mu} \tag{4.10}$$

where u_{ch} is the average fluid velocity in the channel.

The friction coefficient in Eq. (4.9) is a function of the Reynolds number in the channel. In Sect. 4.3 correlations have been obtained from parametric sets of CFD simulations using ANSYS® Fluent. Different correlations are used depending whether a fluid channel is transverse or radial. As the friction coefficient multiplied by the Reynolds number has been observed to be a constant, the hydraulic resistance in each channel becomes independent of the mass flow rate. Thus, for a given flow rate, the hydraulic system of equations becomes linear. However, when the temperature field changes the density and the viscosity also changes, thus the hydraulic resistances of each channel need to be re-calculated.

The computation methodology implemented in FluSHELL hydrodynamic model is described in Fig. 4.13.

After the topological model generates the networks, the hydrodynamic model starts. Beforehand, a first guess of the fluid nodal temperatures is needed to determine the hydraulic resistances. Afterwards the Gauss-Band method is applied to compute the pressure in each node of the fluid network and then the mass flow rates through each channel. That mass flow rate distribution data is then inputted to the heat transfer model, where temperatures are computed. New temperatures are thus calculated for each fluid and solid nodes as well as for each fluid branch. These new temperatures return again to the hydrodynamic model where the hydraulic conductances are again re-calculated and this process is repeated and new pressures are computed. If, between two iterations, the maximum relative pressure variation is below a certain user-defined threshold limit, the calculation is considered converged

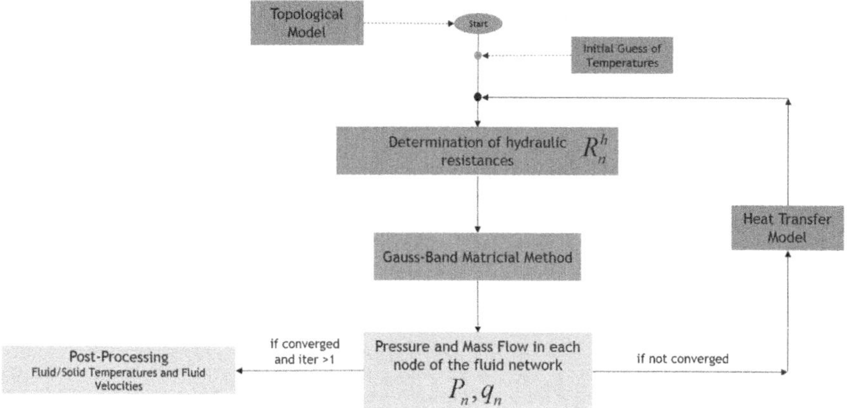

Fig. 4.13 Methodology implemented in FluSHELL to compute the pressures in each node

and the post-processing stage starts. If above that limit, the whole loop is repeated. In the next subsection, the Heat Transfer Model is comprehensively described.

4.2.4　Heat Transfer Model

The Heat Transfer Model comprises the set of equations that describes energy conservation over the fluid and solid networks, considering that the heat generated in the solid is entirely transferred to the fluid and removed from the system, or in that the increase in the fluid enthalpy balances the heat transferred from the solid to the fluid. To express this thermal equilibrium, the model considers two fundamental heat transfer mechanisms: thermal conduction and thermal convection. Thermal conduction is used to describe the heat transferred between solid components of the coil and thermal convection is used to describe the heat transferred between solid and fluid components of the coil.

The result of this model is the temperature in each node of the solid network and the temperature in each node and branch of the fluid network. Both networks are coupled because the temperature in the center of each segment of a turn depends on the nodal temperatures and velocities in the adjacent fluid channels.

First it is necessary to identify the degrees of freedom or paths through which the heat can be transferred. For this purpose, a cut view of a typical coil/washer system is shown in Fig. 4.14 where each component is identified.

The heat is generated inside the *Copper Conductors* represented in Fig. 4.14. These conductors are typically covered by an *Insulation Paper* (generally cellulose Kraft paper). In Fig. 4.14, the internal arrangement of copper conductors is being simplified using an orange bulk material and the external layer is represented using a thinner brown colored material. Each insulated copper conductor (or arrangement

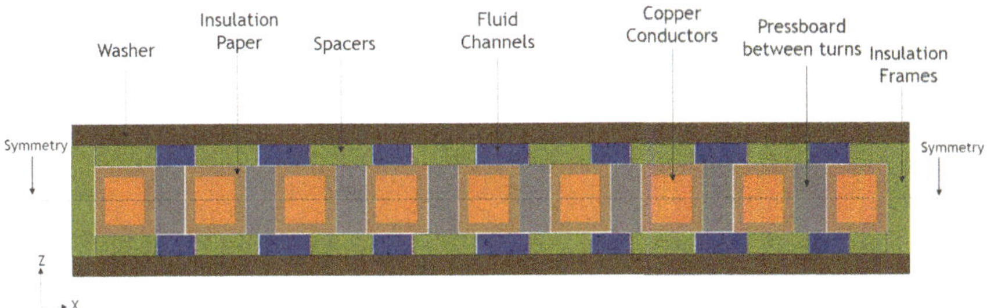

Fig. 4.14 A cut-view (X-Z plane) of a typical coil/washer system. Schematic representation of the main components

of several copper conductors) is said to be a turn. *Pressboard* strips can exist between each two consecutive turns. Moreover, *Insulation Frames* typically exist folded around the innermost and outermost turns (those represented in Fig. 4.14 are U-shaped because they cover both cooling sides of the turn). Thus, all these components are sandwiched between two *Washers* with *Spacers* that in turn create the necessary *Fluid Channels*. All these components together form a coil/washer system. The FluSHELL tool models the thermal-hydraulic behavior of this system.

In steady-state, each copper conductor (or the bulk material representing a group of copper conductors) is a uniform volumetric heat source. Then, at each segment of each turn, heat may be transferred and/or received along each of the six Cartesian directions ($+X$; $-X$; $+Y$; $-Y$; $+Z$; $-Z$), according to

$$Q_{i,j}^g = Q_{i,j,-X} - Q_{i,j,+X} + Q_{i,j,+Y} - Q_{i,j,-Y} + Q_{i,j,+Z} - Q_{i,j,-Z} \qquad (4.11)$$

where $Q_{i,j}^g$ represents the heat generated in segment i, j of a turn, where subscript i refers the turn itself. Each turn in the coil is numbered from $i = 1$, the innermost turn, up to $i = nt$ the outermost turn, and as each turn is then subdivided into segments, subscript j refers the cumulative number of that segment in that turn (counting from the inlet region towards the outlet).

In commercial shell-type coils, the spatial distribution of the spacers of the two washers is similar on either cooling sides of the coil, hence resulting in identical fluid channels on both sides and thus symmetry is assumed, as shown in Fig. 4.14. Accordingly, the Heat Transfer Model assumes that the heat transferred along the $+Z$ direction equals the heat transferred along the $-Z$ direction. Thus, in the center of each turn segment, the heat transferred is null.

Along the $+X$ and the $-X$ directions heat is transferred by thermal conduction between consecutive turn segments as shown in Fig. 4.15.

The innermost turn, $i = 1$, and the outermost turn, $i = nt$, are exceptions because their location at the extents of the system. For the turn segments of the innermost turn the heat transferred along the $-X$ direction is null and for the turn segments of outermost turn the heat transferred along the $+X$ direction is also null.

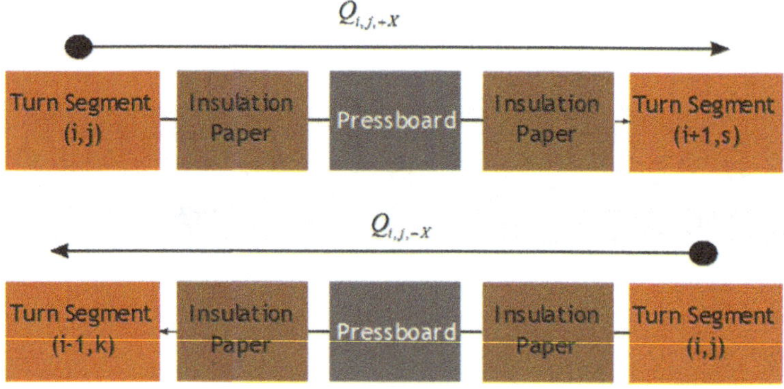

Fig. 4.15 Heat transfer along the +X and −X directions. Identification of components

Along the $-Z$ direction heat is transferred from the turn segments to the corresponding fluid channels across three possible paths as shown in Fig. 4.16. Heat can be either transferred directly across the external insulation paper to the fluid, across the insulation frames to the fluid and across the spacers to the fluid.

Along the $-Y$ and $+Y$ directions heat is transferred between turn segments of the same turn as shown in Fig. 4.17.

The first and last turn segments, $j = 1$, and $j = ns(i)$, of each turn i are exceptions, since in these locations the coil is considered symmetric and thus $Q_{i,1,-Y}$ and $Q_{i,ns(i),+Y}$ are null, respectively.

Heat transfer between node i,j and the neighbouring node in the $+X$ direction described by

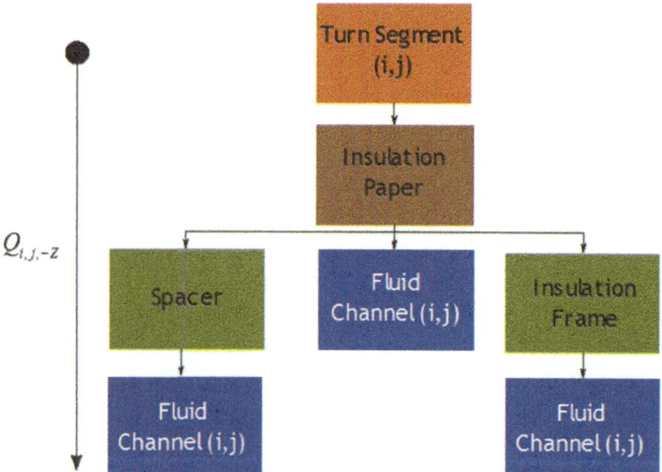

Fig. 4.16 Heat transfer along the −Z direction. Identification of components

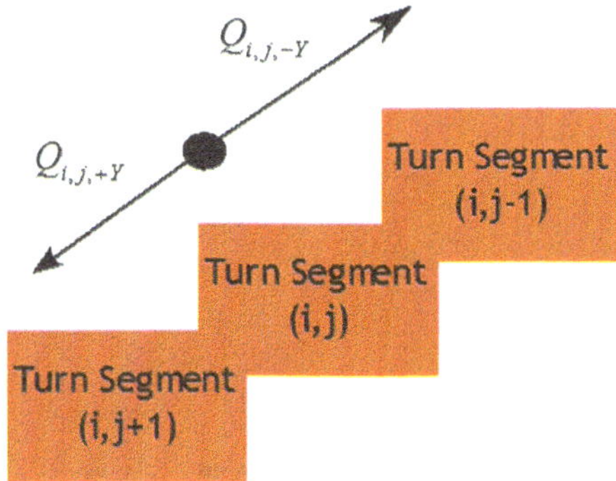

Fig. 4.17 Heat transfer along the +Y and −Y directions. Identification of components

$$\Delta T_{i,j,+X} = R^{t}_{i,j,+X}\, Q_{i,j,+X} \tag{4.12}$$

where $R^{t}_{i,j,+X}$ is the thermal resistance along that direction, or in terms of the respective thermal conductance the equation is re-arranged as

$$Q_{i,j,+X} = \Delta T_{i,j,+X}\, C^{t}_{i,j,+X} \tag{4.13}$$

The thermal resistance $R^{t}_{i,j,+X}$ is obtained by summing the thermal resistances of each individual component, c, that the heat flow crosses from node i,j to a neighbouring node along the $+X$ direction

$$R^{t}_{i,j,+X} = \sum_{c=1}^{nc} R^{c}_{i,j,+X} \tag{4.14}$$

The components along the $+X$ direction have been identified in Fig. 4.15. In Fig. 4.18 these components are represented together with the analogous circuit of thermal resistances. The circuit along the $-X$ direction is identical.

According to Fig. 4.18, the thermal resistance between two neighbouring turn segments i,j and $i+1,s$ along the $+X$ direction is described by the following association of resistances in series

$$R^{t}_{i,j,+X} = R^{bulk}_{i,j,+X} + R^{ext}_{i,j,+X} + R^{pb}_{i,j,+X} + R^{ext}_{i+1,s,+X} + R^{bulk}_{i+1,s,+X} \tag{4.15}$$

where:

- $R^{bulk}_{i,j,+X}$ is the equivalent thermal resistance of the bulk of the turn segment i,j. In the geometry used in this work, the bulk of the turn segment is composed of a single copper conductor, but in most commercial shell-type transformers this equivalent thermal resistance is an association of internal resistances

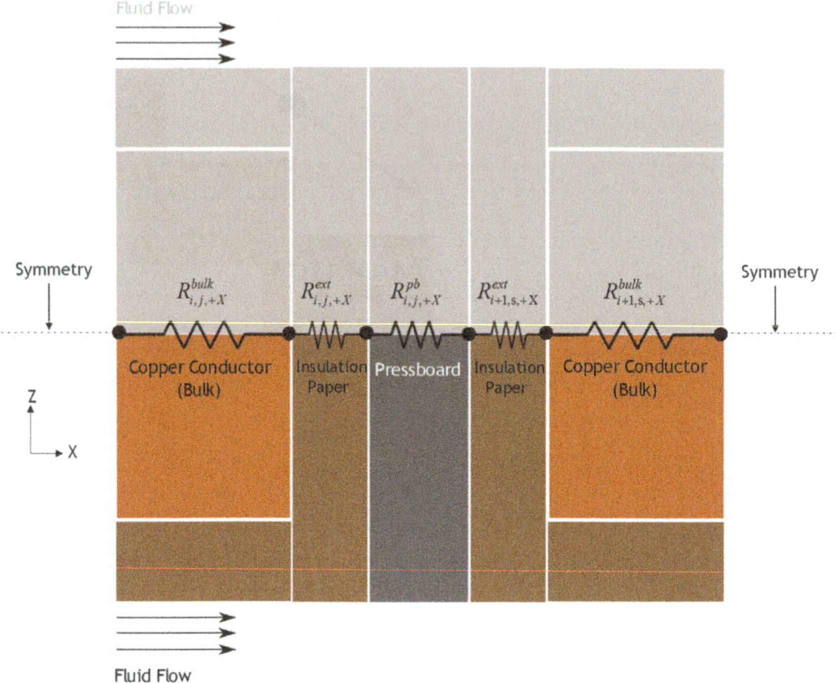

Fig. 4.18 Analogous circuit along the +X direction between two nodes located in the centre of neighbouring turn segments

corresponding to different arrangements of multiple copper conductors. This resistance is obtained by

$$R^{bulk}_{i,j,+X} = \frac{L^{bulk}_{i,j,+X}}{k^{bulk}_{i,j,+X} A^{bulk}_{i,j,+X}} \tag{4.16}$$

where $L^{bulk}_{i,j,+X}$ is the characteristic length of the turn segment i,j, $k^{bulk}_{i,j,+X}$ is its equivalent thermal conductivity and $A^{bulk}_{i,j,+X}$ is the correspondent heat transfer area.

- $R^{bulk}_{i+1,s,+X}$ is the equivalent thermal resistance of the bulk of the neighbouring turn segment $i+1,s$. In the geometry used in this work the turns have the same arrangement of copper conductors and are made of the same materials. This is the case for most commercial shell-type transformers. Thus, this resistance is also computed using Eq. (4.14).
- $R^{ext}_{i,j,+X}$ is the thermal resistance of the external insulation paper covering the turn segment i,j. This resistance is obtained by

$$R^{ext}_{i,j,+X} = \frac{L^{ext}_{i,j,+X}}{k^{ext}_{i,j,+X} A^{ext}_{i,j,+X}} \qquad (4.17)$$

where $L^{ext}_{i,j,+X}$ is the characteristic length of the insulation paper, $k^{ext}_{i,j,+X}$ is its thermal conductivity and $A^{ext}_{i,j,+X}$ is the correspondent heat transfer area.

- $R^{pb}_{i,j,+X}$ is the thermal resistance associated with the pressboard strips existing between each two neighbouring turn segments. In the geometry used in this work pressboard strips with a uniform thickness exist in between each two consecutive turns. This resistance is obtained by

$$R^{pb}_{i,j,+X} = \frac{L^{pb}_{i,j,+X}}{k^{pb}_{i,j,+X} A^{pb}_{i,j,+X}} \qquad (4.18)$$

where $L^{pb}_{i,j,+X}$ is the characteristic length of the pressboard strips between neighbouring turn segments, $k^{pb}_{i,j,+X}$ is its thermal conductivity and $A^{pb}_{i,j,+X}$ is the correspondent heat transfer area.

The components along the $-Z$ direction have been identified in Fig. 4.16. In Fig. 4.19 these components are represented together with the analogous circuit of thermal resistances. The circuit along the $+Z$ direction is identical; however the coil/washer is assumed symmetric along this reverse direction.

The thermal resistance between a turn segment and the corresponding adjacent fluid channel depends upon whether that segment is in direct contact with the fluid or with a solid (either spacer or insulation frame) before the fluid (Fig. 4.16). Both cases are addressed below.

In Fig. 4.19 the turn segment i,j is in direct contact with a fluid channel and its equivalent thermal resistance along the $-Z$ direction is described by the following association of resistances in series

$$R^{t}_{i,j,-Z} = R^{bulk}_{i,j,-Z} + R^{ext}_{i,j,-Z} + R^{fluid}_{i,j,-Z} \qquad (4.19)$$

where:

- $R^{bulk}_{i,j,-Z}$ is the equivalent thermal resistance of the bulk of the turn segment i,j. This resistance is obtained using an expression like Eq. (4.16).
- $R^{ext}_{i,j,-Z}$ is the thermal resistance of the external insulation paper covering the turn segment i,j. This resistance is obtained using an expression like Eq. (4.17).
- $R^{fluid}_{i,j,-Z}$ is the thermal resistance of the adjacent fluid channel. This resistance is obtained by

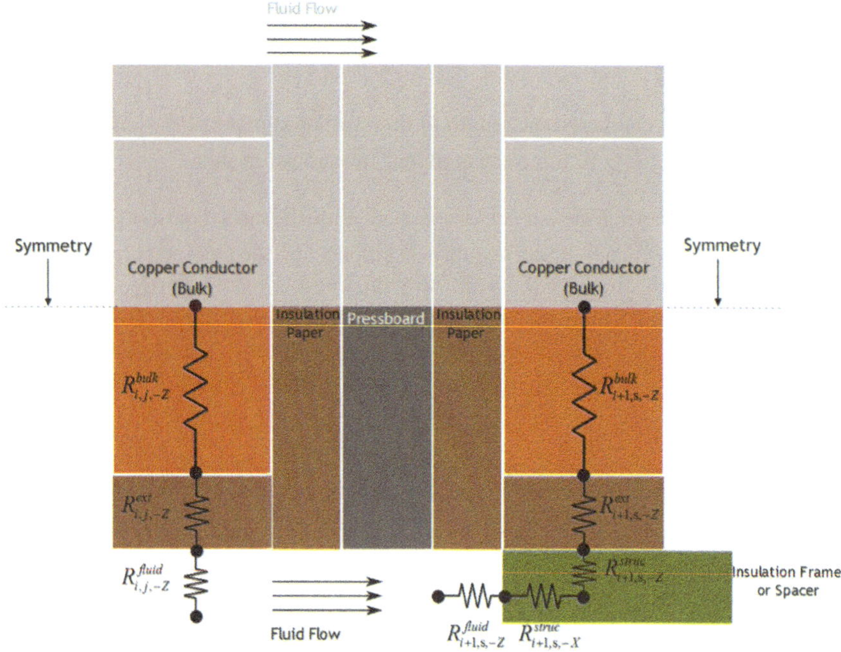

Fig. 4.19 Analogous circuits between nodes in the centre of the turn segments and the corresponding nodes in the fluid channels (along the –Z direction)

$$R_{i,j,-Z}^{fluid} = \frac{1}{U_{i,j,-Z}^{fluid} A_{i,j,-Z}^{fluid}} \tag{4.20}$$

where $U_{i,j,-Z}^{fluid}$ is the heat transfer coefficient of the fluid channel and $A_{i,j,-Z}^{fluid}$ is the area of the turn segment i,j wetted by the fluid. As for many other practical engineering applications, an analytical derivation of these heat transfer coefficients demands a mathematical model to describe the temperature profile near the turn segment surfaces. Instead, in FluSHELL, these coefficients are computed using correlations that have been in turn obtained from parametric sets of 3D CFD simulations. These simulations and corresponding correlations are described in Sect. 4.3.

This heat transfer coefficient depends upon the average fluid velocity in the adjacent fluid channel as well as its temperature. Thus, reinforcing the statement in the beginning of this section that both fluid and solid networks are coupled.

In Fig. 4.19 the turn segment $i+1, s$ is in direct contact either with a spacer or with an insulation frame and its thermal resistance along the $-Z$ direction is described by the following association of resistances in series

$$R^t_{i+1,s,-Z} = R^{bulk}_{i+1,s,-Z} + R^{ext}_{i+1,s,-Z} + R^{struc}_{i+1,s,-Z} + R^{struc}_{i+1,s,-X} + R^{fluid}_{i+1,s,-Z} \qquad (4.21)$$

where the first three terms are similar to Eq. (4.19) and the two additional resistances are due to either the spacer or the insulation frame existing between the insulation paper and the fluid channel:

- $R^{struc}_{i+1,s,-Z}$ is the thermal resistance of either the spacer or the insulation frame along the $-Z$ direction. This resistance is obtained by

$$R^{struc}_{i+1,s,-Z} = \frac{L^{struc}_{i+1,s,-Z}}{k^{struc}_{i+1,s,-Z} A^{struc}_{i+1,s,-Z}} \qquad (4.22)$$

where the characteristic length, $L^{struc}_{i,j,-Z}$, corresponds to the half thickness of the spacers or the half thickness of the insulation frames.

- $R^{struc}_{i+1,s,-X}$ is the thermal resistance of either the spacer or the insulation frame along the perpendicular $-X$ direction. This resistance is obtained by

$$R^{struc}_{i+1,s,-X} = \frac{L^{struc}_{i+1,s,-X}}{k^{struc}_{i+1,s,-X} A^{struc}_{i+1,s,-X}} \qquad (4.23)$$

where the characteristic length, $L^{struc}_{i+1,s,-X}$, corresponds to the distance between the centroid of the spacers or of the insulation frame and the fluid channel.

Along the $+Y$ and the $-Y$ directions the heat is transferred exclusively across the bulk of the turn segment, resulting in the following thermal resistance

$$R^t_{i,j,-Y} = R^{bulk}_{i,j,-Y} \qquad (4.24)$$

where $R^{bulk}_{i,j,-Y}$ is the equivalent thermal resistance of the bulk of the turn segment i,j along the $-Y$ direction. The thermal resistance along the reversed direction $+Y$ is identical and the expression describing this thermal equivalent resistance is like Eq. (4.16).

The energy balance to a generic turn segment i,j is depicted in Fig. 4.20.

Figure 4.20a depicts a turn segment i,j receiving heat from its $-X$ neighbouring turn segments $(i-1,k;i-1,nb^*)$ and transferring heat to its $+X$ neighbours $(i+1,s;i+1,nb)$. Along this coordinate heat can be transferred to multiple turn segments depending on the location within the same coil. For instance, the specific segment highlighted in Fig. 4.10c transfers/receives heat from two neighbour turn segments along the $+Y$ direction and transfers/receives heat from a single neighbour turn segment along the $-Y$ direction.

Similarly, this generic turn segment i,j is receiving heat from the turn segment $i,j+1$ along the $+Y$ direction and transferring heat to the turn segment $i,j-1$ along the $-Y$ direction.

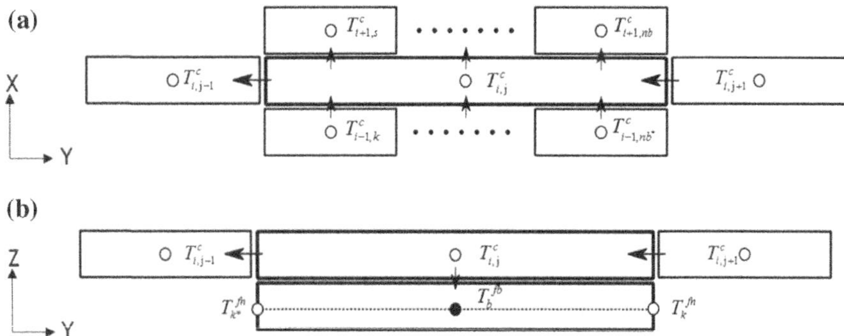

Fig. 4.20 Energy balance to a generic turn segment i, j: **a** along X and Y coordinates and **b** along Z and Y coordinates

In Fig. 4.20b this generic turn segment i, j is transferring heat to a fluid channel b. The fluid channel has a node k representing the inlet and a node k^* representing the outlet. This fluid channel might be in direct contact with the generic segment i, j or might be connected to it either through a spacer or an insulation frame (as referred in Fig. 4.19). The difference between both cases is the thermal resistance, as described in Eq. (4.19) and in Eq. (4.21).

Finally, the heat transferred and/or received along the three Cartesian coordinates must balance the heat being generated in each turn segment. Thus, writing Eq. (4.11), in order to describe the specific energy balance depicted in Fig. 4.20, results in

$$Q_{i,j,+Y} + \sum_{k=1}^{nb*} Q_{i,j,k} = \sum_{s=1}^{nb} Q_{i,j,s} + Q_{i,j,-X} + Q_{i,j,-Y} + Q_{i,j,-Z} + Q_{i,j}^g \tag{4.25}$$

where nb and nb^* refer to the number of neighbouring nodes in the two opposing directions from the centre of each turn segment.

Substituting Eq. (4.13) into (4.25) results in

$$\frac{T_{i,j+1}^c}{R_{i,j,+Y}^t} - \frac{T_{i,j}^c}{R_{i,j,+Y}^t} + \sum_{k=1}^{nb*}\left(\frac{T_{i-1,k}^c}{R_{i,j,-X}^t}\right) - \frac{T_{i,j}^c}{\sum\limits_{k=1}^{nb*}\left(R_{i,j,-X}^t\right)} =$$

$$= \frac{T_{i,j}^c}{\sum\limits_{s=1}^{nb}\left(R_{i,j,+X}^t\right)} - \sum_{s=1}^{nb}\left(\frac{T_{i+1,s}^c}{R_{i,j,+X}^t}\right) + \frac{T_{i,j}^c}{R_{i,j,-Y}^t} - \frac{T_{i,j-1}^c}{R_{i,j,-Y}^t} + \frac{T_{i,j}^c}{R_{i,j,-Z}^t} - \frac{T_k^{fn}}{R_{i,j,-Z}^t} + Q_{i,j}^g$$

$$\tag{4.26}$$

where: $T_{i,j}^c$ represents the temperature in the node of turn segment i, j, where the balance is being expressed; $T_{i,j-1}^c$ and $T_{i,j+1}^c$ represent the temperatures in nodes located in neighbouring turn segments within the same turn along the $-Y$ and $+Y$ directions; $T^c_{i-1,k}$ and $T^c_{i+1,s}$ represent the temperatures in the solid nodes located in the $-X$ and $+X$ neighbour segments; T^{fn}_k represents the temperature in the fluid node located at the inlet of the fluid channel associated with the turn segment i,j; and $Q^g_{i,j}$ represents the heat generated in this turn segment.

Re-arranging previous Eq. (4.26) results in

$$T^c_{i,j} + \frac{a_{i,j+1}}{a_{i,j}} T^c_{i,j+1} + \frac{a_{i,j-1}}{a_{i,j}} T^c_{i,j-1} + \sum_{k=1}^{nb^*} \left(\frac{a_{i-1,k}}{a_{i,j}} T^c_{i-1,k} \right) + \sum_{s=1}^{nb} \left(\frac{a_{i+1,s}}{a_{i,j}} T^c_{i+1,s} \right) = b_{i,j} T^{fn}_k + \frac{Q^g_{i,j}}{a_{i,j}}$$

$$(i = 1\ldots nt; j = 1\ldots ns(i))$$

$$(4.27)$$

where nt is the total number of turns and $ns(i)$ is the total number of segments in each turn. And:

$$a_{i,j} = \left(-\frac{1}{R^t_{i,j,+Y}} - \frac{T^c_{i,j}}{\sum\limits_{k=1}^{nb^*} \left(R^t_{i,j,-X} \right)} - \sum_{s=1}^{nb} \left(\frac{1}{R^t_{i,j,+X}} \right) - \frac{1}{R^t_{i,j,-Y}} - \frac{1}{R^t_{i,j,-Z}} \right) \quad (4.28)$$

$$a_{i,j+1} = \frac{1}{R^t_{i,j,+Y}} \quad (4.29)$$

$$a_{i,j-1} = \frac{1}{R^t_{i,j,-Y}} \quad (4.30)$$

$$a_{i-1,k} = \frac{1}{R^t_{i,j,-X}} \quad (4.31)$$

$$a_{i+1,s} = \frac{1}{R^t_{i,j,+X}} \quad (4.32)$$

$$b_{i,j} = \frac{1}{a_{i,j} R^t_{i,j,-Z}} \quad (4.33)$$

Then the application of the energy balance in Eq. (4.27) to each segment of each turn, results in a system of non-linear equations. This non-linear system of algebraic equations is solved iteratively using the Fixed-Point method.

Finally, the set of required equations is closed by writing the energy balances to the fluid. A schematic representation is shown in Fig. 4.21.

The energy balance for fluid node k depicted in Fig. 4.21a is given by

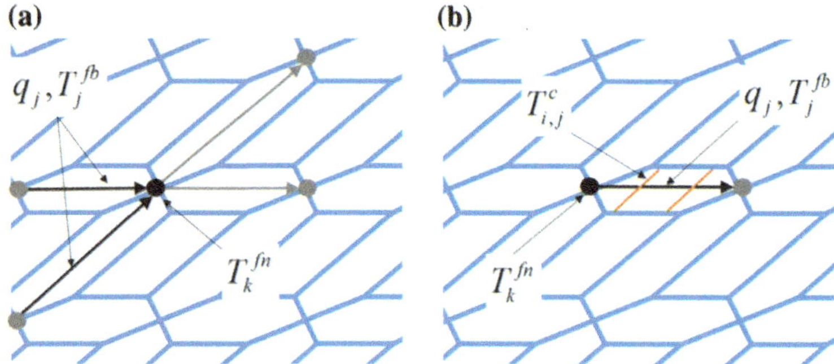

Fig. 4.21 Energy balances on the fluid network: **a** generic fluid node and **b** generic fluid branch

$$\sum_{j=1}^{nconv} Cp\left(T_j^{fb}\right) q_j \, T_j^{fb} = Cp\left(T_k^{fn}\right) T_k^{fn} \sum_{j=1}^{nconv} q_j$$

$$(k = 1 \ldots nfn)$$

(4.34)

where *nfn* is the total number of fluid nodes; *nconv* represents the number of branches connected to node k; $Cp\left(T_j^{fb}\right)$ corresponds to the fluid specific heat capacity evaluated at the temperature, T_j^{fb}, of each connected branch j; $Cp\left(T_k^{fn}\right)$ corresponds to the fluid specific heat capacity evaluated at the temperature, T_k^{fn}, of node k; and q_j the mass flow rate through branch j.

The energy balance for fluid branch j depicted in Fig. 4.21b is given by

$$Cp\left(T_k^{fn}\right) T_k^{fn} q_j + \sum_{i,j} \left(U_{i,j,-z}^{fluid} A_{i,j,-z}^{fluid} \left(T_{i,j}^c - T_k^{fn}\right) \right) = Cp\left(T_j^{fb}\right) T_j^{fb} q_j$$

$$(j = 1 \ldots nfb)$$

(4.35)

where *nfb* is the total number of fluid branches; $T_{i,j}^c$ represents the temperature of the turn segments in contact with branch j; $A_{i,j,-z}^{fluid}$ represents the respective heat transfer area of those segments; and $U_{i,j,-z}^{fluid}$ is the heat transfer coefficient of the different segments contacting with branch j.

The temperatures of each fluid node and fluid branch are then calculated using Eqs. (4.34) and (4.35), respectively.

The solution of the systems of equations used in this model is obtained iteratively as shown in Fig. 4.22.

The heat transfer model imports from the hydrodynamic model both the mass flow rate distribution in each channel and, in the case of the first iteration, also imports the initial temperature guesses. Thus, the procedure comprised three main

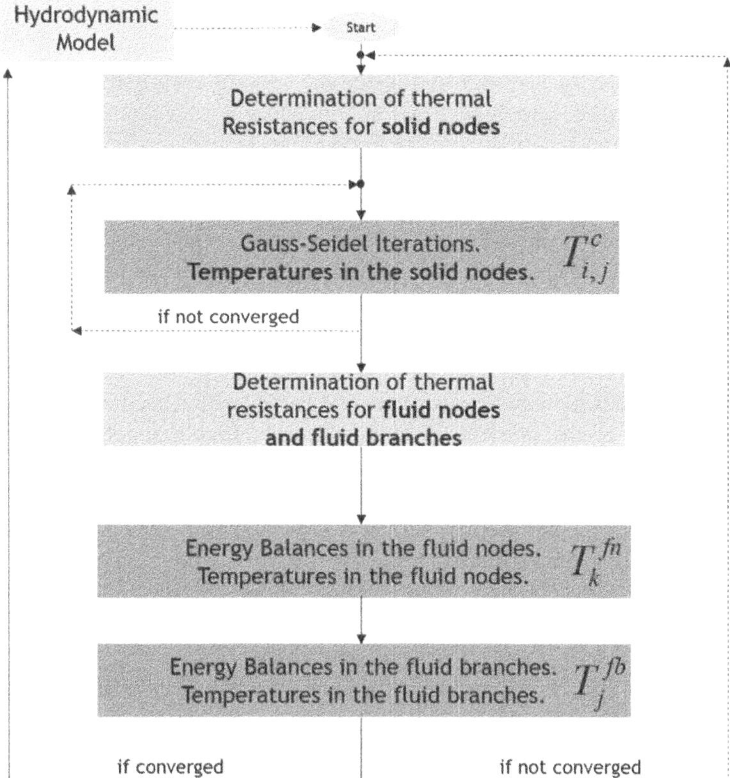

Fig. 4.22 Methodology implemented in FluSHELL to compute the temperatures in each node (both solid and fluid) and in each branch

steps with the aim of computing the temperatures in the solid nodes, in the fluid nodes and in the fluid branches.

First, the equations for the solid temperature nodes—Eq. (4.27)—are solved iteratively using Gauss-Seidel method until a certain converge criterion is observed. This enables the calculation of temperatures in each turn segment, $T^c_{i,j}$.

Second, the additional energy balances in the fluid nodes are solved followed by the energy balances in the fluid branches. Equations (4.34) and (4.35), respectively. This enables the calculation of the fluid temperatures in the nodes, T^{fn}_k, and the fluid temperatures in the branches, T^{fb}_j.

If the maximum relative temperature variation is below a certain user-defined threshold limit, the temperature distribution is considered converged and inputted again to the hydrodynamic model. Finally, the hydrodynamic model controls the decision whether or not consider the whole calculation converged, depending on the pressure variation from iteration to iteration.

4.3 FluSHELL Calibration

The flow regime of the cooling fluid inside the coils of shell-type transformers is laminar and the most commonly used cooling fluid is naphthenic mineral oil, a Newtonian fluid under normal operating conditions (Langhame et al. 1985).

This enables the CFD computation of the velocity and temperature profiles near the walls without simplifications, which means the CFD can compute fluid velocities and fluid temperatures up to the hot walls, without demanding further simplifications. However, in the FluSHELL tool these temperature profiles are approximated by lumped coefficients: the friction coefficient, f, identified in Eq. (4.9), represents the fluid velocity profile and the heat transfer coefficient, U^{fluid}, identified in Eq. (4.20), represents the fluid temperature profile. The determination of the friction coefficients of the hydraulic resistances within each channel and thus the computation of the pressure drop inside each fluid channel using a single average fluid velocity, while the determination of the heat transfer coefficients enable the computation of the heat removed from each turn segment, using a single temperature difference between its surface temperature and the bulk fluid temperature in the adjacent fluid channel.

These assumptions are the main reason behind the lower time-to-solution exhibited by thermal-hydraulic network tools, such as FluSHELL, when compared with CFD tools.

Therefore, FluSHELL accuracy is dependent on these coefficients and hence an adequate methodology to obtain such coefficients is paramount. This has been the main objective of recent works that used CFD to obtain improved friction and heat transfer correlations for the windings of typical core-type transformers (Campelo et al. 2012; Coddé et al. 2015; Wu et al. 2012, b). In two of these works the authors conducted sets of CFD simulations in a theoretical T-Junction configuration in order to obtain improved correlations for the frictional losses and heat transfer coefficients (Wu et al. 2012a, b). This assumes relevance when the flow and geometric assumptions underlying expressions available in literature deviate from the conditions occurring in a commercial transformer. In one of the works the authors conducted the CFD simulations with the goal of extracting junction pressure losses coefficients. Because the network model with literature expressions yielded strong unrealistic pressure losses (Wu et al. 2012).

This section describes an identical process of extracting friction and heat transfer coefficients from CFD simulations, now applied for the first time to shell-type transformers.

This section begins by describing the CFD model used to extract the data Sect. 4.3.1. Afterwards the friction and heat transfer coefficients are described in the Sect. 4.3.2. Part of this work has been reported in a previous master thesis co-supervised by the author (Oliveira 2014), however this section contains new results. These results have been partially reproduced in two conference papers (Campelo et al. 2014b, 2015b).

4.3.1 CFD Model

4.3.1.1 Geometry

Modelling a complete electrical transformer cooling loop using CFD is not common due to the complex multiscale computational domain that needs to be considered; so it is imperative to restrain the computational domain to the smallest representative region. In Shell-Type transformers the smallest representative region of a winding is the coil/washer system. In this system, the fluid velocity and temperature profiles near the walls cannot be adequately modelled using 2D planes, as is the case for core-type windings where most of the CFD analysis are usually conducted under 2D axisymmetric conditions and then extrapolated to 3D conditions (Campelo et al. 2009; Torriano et al. 2010). In this case the CFD simulations are necessarily 3D.

This coil/washer system is shown in Fig. 4.23. This system implies modelling in 3D the turns, the fluid channels, the spacers and the additional insulation frames located at the borders (if they exist).

Despite this, due to similar arrangements of fluid channels in both cooling sides of each coil, the coil/washer system can be further simplified by applying symmetry at half height of the turns (*XY* plane dotted in Fig. 4.23a) and by applying a longitudinal symmetry on the washer (*XZ* plane dotted in Fig. 4.23b). In this case, for obtaining the friction and heat transfer coefficient correlations, the turns of the coil have been approximated by a constant heat flux wall (Fig. 4.24).

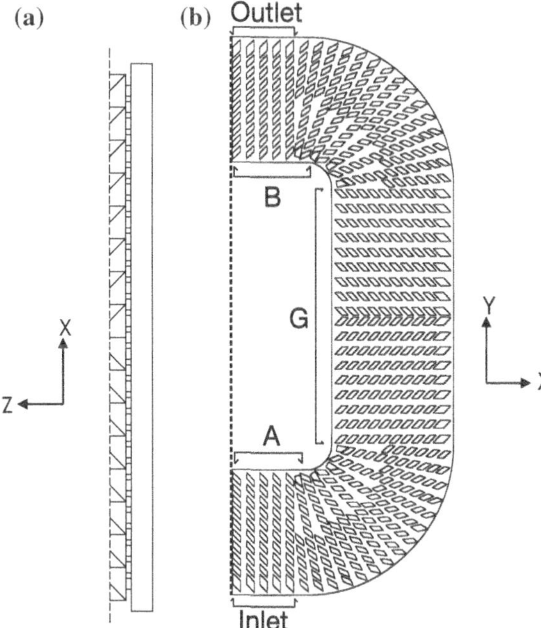

Fig. 4.23 Smallest representative 3D domain. **a** XZ plane with symmetry plane at half height of the turns, **b** YX plane with longitudinal symmetry

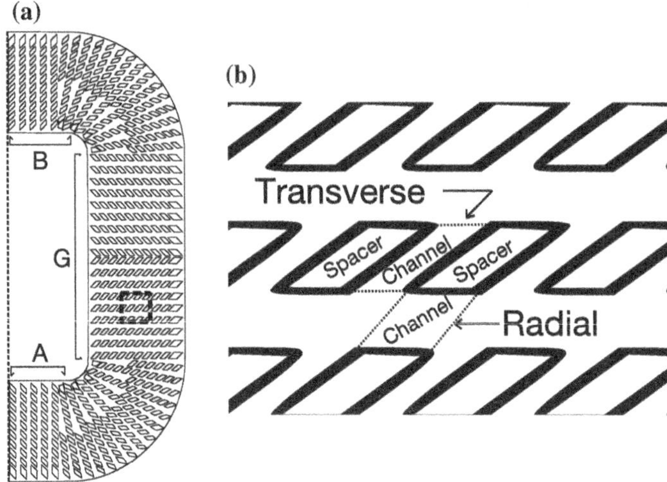

Fig. 4.24 Identification of the two types of fluid channels considered (transverse and radial): **a** Location to be zoomed and **b** zoomed location

Table 4.3 summarizes the reference dimensions of the computational domain. This computational domain is quite similar to the scale model described in Chap. 2. In this case, there a few spacers with different shapes namely in the inner part of the curved sections of the washers. The scale model geometry has been further simplified to have all the spacers with the same dimensions. The labels A, B and G refer to different sections of the coil/washer system as shown in Fig. 4.25.

As shown in Fig. 4.23 the inlets are in the bottom of the domain (Y = 0 m), between the symmetry YZ plane and the fifth vertical column of spacers. The outlets are symmetrically located at the top of the domain.

The cooling fluid flow path geometry is mainly constituted by fluid channels opened by spacers. There are channels between consecutive spacers of the same row—Radial Fluid Channels—and channels between consecutive rows of spacers —Radial Fluid Channels. Both types of channel are identified in Fig. 4.24.

Table 4.3 Reference dimensions (in m) of the computational domain used for calibration	Location	Coordinate	Dimensions [m]
	A = B	Y	0.276
		X	0.170
	G	Y	0.561
		X	0.276
	Distance between spacers	Y	0.015
		X	0.008
	Height of spacers	Z	0.00195

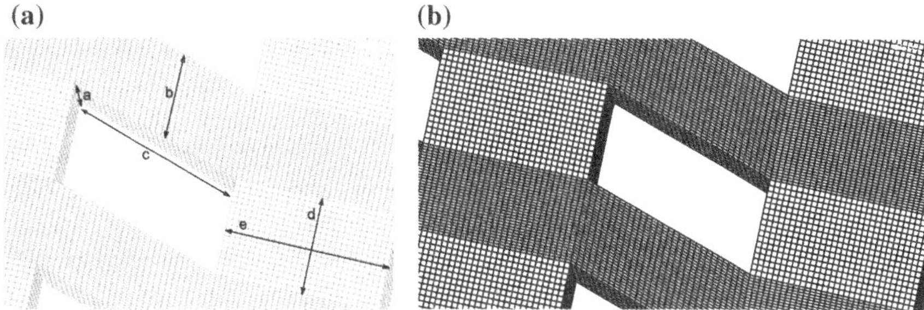

Fig. 4.25 Mesh used in the sensitivity analysis: **a** main mesh directions **b** mesh volumes used in the radial and transverse fluid channels

These transverse and radial fluid channels resent distinct geometrical dimensions and associated characteristics. For instance, the transverse channels have (on average) a lower aspect ratio than the radial channels: 4.3 and 7.5, respectively. In turn the average length of the transverse and radial channels is 23.3 and 13.3, respectively. Table 4.4 summarizes the main hydraulic and thermal characteristics of these channels. For this purpose, the data in Table 4.4 refers to the sample fluid channels created to extract transport properties from the CFD according to Fig. 4.35.

For hydraulically fully developed flow between infinite parallel plates the Darcy friction coefficient can be analytically demonstrated to be $f = 96/Re$ (Shah and London 1971). As flow develops the friction coefficient progress towards this fully developed value. Based on the data provided by (Roshenow et al. 1998) the flow becomes fully developed at $x^h = 0.015Re \times d_h$. Consequently, this expression has

Table 4.4 Main characteristics of transverse and radial channels using data extracted from the sample fluid channels (data from valid channels)

	Transverse channels	Radial channels
Height [mm]	1.95	1.95
Average length [mm]	23.3	13.3
Average width [mm]	8.3	14.7
Average aspect ratio (Width/Height)	4.3	7.5
Reynolds range (Min.—Max.)	4–176	9–223
Average Reynolds	76	55
Hydraulic length (Min.-Max.) [mm]	0.2–10.3	0.5–13.1
Average hydraulic Length [mm]	4.5	3.2
Hydraulic length (Min.-Max.) [%]	1–44	4–98
Average hydraulic Length [%]	19	24
Thermal length (Min. - Max.) [mm]	21–1030	50–1305
Average thermal length [mm]	446	323
Thermal length [%]	90–4413	215–5594
Average thermal length [%]	1913	1383

been used to estimate the average hydraulic length, x^h, of the transverse and radial channels shown in Table 4.4. On the other hand, the thermal hydraulic entrance length has been estimated as the product of the hydraulic thermal length and the Prandtl Number of the cooling fluid, $x^t = x^h Pr$ (with $Pr \approx 100$ for the naphthenic mineral oils used in transformers).

From the characteristics in Table 4.4, the flow is then expected to be fully developed along 81% of the length in the transverse channels and along 76% of the length in the radial channels. There are exceptions, but this is the expected average pattern. However, due a high value of the Prandtl Number, the thermal entrance length required can be of the order of meters in both transverse and radial channels, meaning that this distance exceeds largely the average length available in both types of channels.

As a result, the heat transfer coefficients are expected to be dependent on the Reynolds number. The potential deviation from ideal flow and thermal characteristics, underpinned a verification process with the aim of assessing and calibrating transport properties as the friction coefficients and heat transfer coefficients.

For this purpose, a 3D CFD model has been built using the smallest representative region identified in Fig. 4.24. The mesh, the boundary conditions and the main results from this 3D CFD model are described along in the next subsections.

4.3.1.2   Mesh

A 3D model has been built in ANSYS Workbench® 15.0. The mesh used, shown in Fig. 4.25, is hexahedral and structured in most of the computational domain particularly in the fluid channels used to further extract the relevant transport properties.

As depicted in Fig. 4.25, the mesh volumes used are uniformly distributed with no increase of layers near the walls. The spacers (white coloured in Fig. 4.25) are not meshed and the respective walls in contact with the surrounding fluid are considered adiabatic. The coils are modelled using a constant heat flux wall (*hot plate*) as shown schematically in Fig. 4.26.

Before proceeding to extract friction and heat transfer coefficients, a prior mesh sensitivity analysis has been conducted to assess numerical uncertainties.

Table 4.5 describes the group of different mesh sizes used in the sensitivity analysis indicating the number of elements along the main mesh directions identified in Fig. 4.25a, the total number of mesh volumes and few mesh quality parameters.

Fig. 4.26 Schematic representation of the equivalent constant heat flux wall (hot plate) used to model the coil

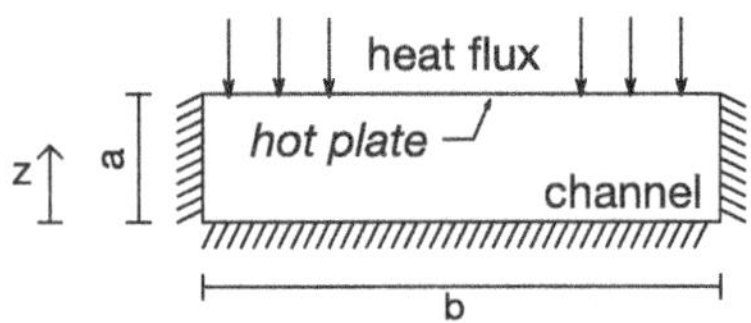

Table 4.5 Characteristics of the 3 mesh sizes used for the sensitivity analysis

Direction	Mesh volumes per direction		
	Mesh6	Mesh9	Mesh12
a	6	9	12
b	31	31	31
c	30	30	30
d	17	17	17
e	30	30	30
Total number of mesh volumes	4 415 826	6 218 739	8 291 652
Mesh quality parameters			
Hot plate nearest centroid (m)	0.0001625	0.0001083	0.0000813
Minimum orthogonal quality	0.55	0.55	0.55
Maximum mesh volume aspect ratio	6.96	10.38	13.81

The three mesh sizes present similar values except the number of mesh volumes along direction a, that is Mesh6, Mesh9 and Mesh12 with 6, 9 and 12 elements, respectively. The nearest hot plate centroid distance progressively decreases from 0.1625 mm to 0.0813 mm, hence increasing the capability to compute fluid velocity and temperature profiles near the walls more accurately.

CFD simulations have been conducted in ANSYS Fluent® 15.0 using its pressure based solver under laminar conditions. A coupled scheme for the pressure-velocity has been used along with a Courant Flow number of 200.

Table 4.6 summarizes the main boundary conditions imposed in the CFD simulations for the mesh sensitivity analysis.

A constant total pressure of 800 Pa has been imposed at the inlet region and a static pressure of 0 Pa at the outlet region.

The reference cooling fluid is Nynas Taurus®. All physical properties have been considered temperature dependent and most of the values used were obtained from the supplier, except the dynamic viscosity which has been determined in a previous work (Gomes et al. 2007b). The physical properties of the cooling fluid used in the in CFD model are listed in Table 4.7.

Continuity, energy and momentum equations for the three Cartesian components of the velocity have been computed. Numerical residuals of 10^{-6} for continuity and momentum equations and residuals of 10^{-9} for the energy equation have been

Table 4.6 Boundary conditions used for the mesh sensitivity analysis

Inlet	Total pressure	800 Pa
	Temperature	60 °C
Outlet	Static pressure	0 Pa
Heat Flux	Constant heat flux	1520 W.m^{-2}
	Momentum	No-slip condition
Other Walls	Momentum	No-slip condition
	Heat transfer	Adiabatic

Table 4.7 Physical properties of the cooling fluid as implemented in CFD

Type of fluid (commercial reference)	Properties
Naphtenic mineral oil (Nynas Taurus®)	Dynamic viscosity, μ [kg.m^{-1}.s^{-1}]
	$1.43276e - 7 \times Exp(3479.5/T(K))$
	Density, ρ [kg.m^{-3}]
	$868 * (1 - 0.00064 \times ((T(K) - 273.15) - 20))$
	Specific heat capacity, Cp [J.kg^{-1}.°C^{-1}]
	$3.4549 \times (T(K) - 273.15) + 1796.2$
	Thermal conductivity, k [W.m^{-1}.°C^{-1}]
	$-7.77e - 5 \times (T(K) - 273.15) + 0.132949$

considered as a condition of convergence but not self-sufficient. Local velocities and temperatures in randomly chosen locations have been also monitored and observed to be varying, between iterations, on the same order of magnitude as the residuals. The global sensitivity results for the three meshes are shown in Table 4.8 using 2nd order discretization schemes.

The average oil temperature as well as at the inlet and outlet oil temperatures represent a volume weighted average, while the maximum temperatures represent the highest values in a single finite volume face. These quantities are listed in Table 4.8 for the three meshes.

Table 4.8 shows temperature differences higher than 2 °C while comparing Mesh6 with highest resolution Mesh12. These differences decrease to less than 0.5 °C when comparing Mesh9 against Mesh12. A difference of less than 3% in the inlet mass flow rate is observed.

Specific fluid channels have been chosen for this study and the corresponding locations are shown in Fig. 4.27.

Each location is identified with two indexes: the first index concerns to the spacer row number immediately preceding the fluid channel and the second index concerns the number of the spacer immediately before the fluid channel (counting from the innermost region of the washer to the outermost). For example, 2–2 refers to a location between the second and the third rows of pressboard spacers and

Table 4.8 Global mesh sensitivity results

	Mesh6	Mesh9	Mesh12
Inlet mass flow rate, q_{inlet} [kg.s^{-1}]	0.00691	0.00682	0.00674
Inlet oil temperature, $T_{oil,1}$ [°C]	60.0	60.0	60.0
Outlet oil temperature, $T_{oil,2}$ [°C]	91.7	93.2	93.3
Average oil temperature, $\bar{T}_{oil}$ [°C]	76.4	77.3	77.2
Maximum oil temperature, $T_{max,oil}$ [°C]	105.2	107.6	108.1
Average hot plate temperature, $\bar{T}_{hotplate}$ [°C]	83.5	84.4	84.5
Maximum hot plate temperature, $T_{max,hotplate}$ [°C]	107.2	109.0	109.4

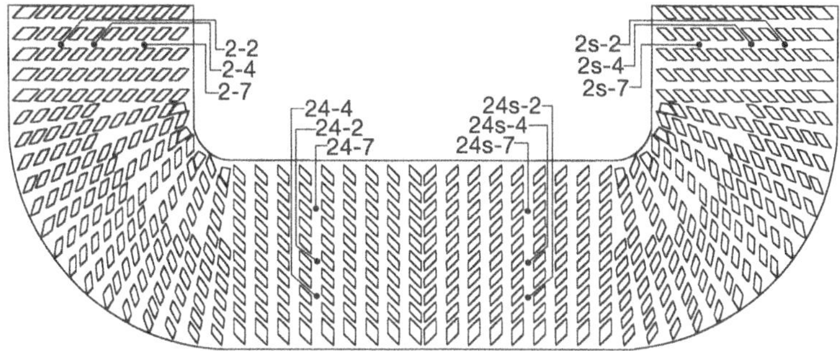

Fig. 4.27 Locations of the fluid channels used to evaluate the mesh sensitivity

between the second and the third pressboard spacer. The locations identified with a 's' after the first index mean that the referred location is symmetric to another location on the left side of the domain (Fig. 4.27).

Each of the 12 locations depicted comprises a pair of fluid channels: one transverse and one radial channel. All the surfaces corresponding to these transverse and radial fluid channels have their own entities that have been created during the pre-processing stage. This has been found to be a relevant best practice in order to avoid additional interpolation errors during the post-processing stage.

The objective of these simulations is to extract transport properties (e.g. friction and heat transfer coefficients) and not to compare against other simulations or measurements. Thus, rather than compare temperatures, the mesh sensitivity has been assessed using quantities that directly influence this methodology of extracting transport properties:

- the shear stress magnitude, used to compute friction coefficients;
- the local temperature difference between the hot plate and the oil temperature at the inlet of each particular fluid channel, used to compute the heat transfer coefficients.

Figure 4.28 depicts the total shear stress magnitude evaluated over the surfaces of each fluid channel. The values are shown as a relative difference (in %) to the highest resolution Mesh12.

Figure 4.28 shows that, for both transverse and radial channels, the absolute relative deviations of the total shear stress magnitude are persistently lower than 2%, even for the coarsest mesh scenario (Mesh6). The friction coefficient is directly proportional to the total shear stress magnitude, hence a maximum influence of this order of magnitude can be expected.

Table 4.9 lists the relative deviations of the average wall temperature difference to the oil entering each channel.

The results in Table 4.9 show an average absolute deviation below 3% for the coarsest Mesh6 while for Mesh9 the average deviation is lower than 2%. The heat transfer coefficient depends on the average wall temperature difference to the oil

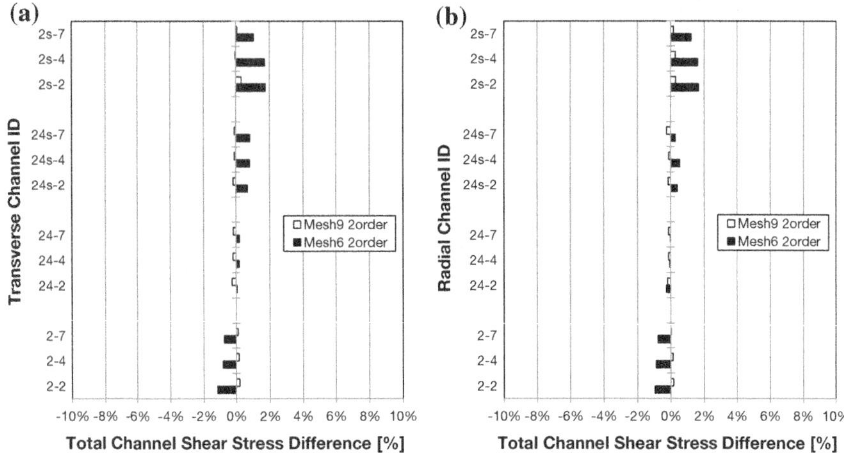

Fig. 4.28 Influence of the mesh size in the total shear stress: **a** transverse channels, **b** radial channels

Table 4.9 Influence of the mesh size in the average wall temperature difference to the oil entering each channel

Channel	Relative deviation to Mesh12 [%]			
	Transverse channels		Radial channels	
	Mesh6	Mesh9	Mesh6	Mesh9
2s-7	−3.5	−1.1	−1.7	−0.4
2s-4	−2.7	−0.6	−1.7	−0.3
2s-2	−2.3	−0.7	−1.4	−0.2
24s-7	−2.0	−0.7	−2.7	−1.0
24s-4	−2.0	−8.2	−3.1	−1.2
24s-2	−2.7	−1.0	−3.1	−1.3
24–7	−2.5	−1.0	−4.4	−1.9
24–4	−2.8	−1.2	−3.6	−1.5
24–2	−3.1	−1.4	−3.2	−1.4
2–7	−2.8	−1.0	−2.3	−1.0
2–4	−2.5	−0.5	−3.3	−1.2
2–2	−3.1	−0.8	−3.3	−0.9
Average Abs. deviation (%)	2.7	1.5	2.8	1.0

entering each channel, hence a maximum influence of these order of magnitude can be expected.

Considering that the resulting mass flow rate also varied on the same order of magnitude, these numerical uncertainties below 3% for the coarsest mesh have been considered adequate for the purpose, and the parametric CFD simulations to extract transport properties have been conducted using Mesh6.

4.3.1.3 Boundary Conditions

The boundary conditions and other parameters of the simulations are listed in Table 4.10.

Different values for the inlet mass flow rates, q_{inlet}, were considered in order to widen the analysis to common design ranges and are listed in Table 4.11.

In Table 4.11 each simulation is identified by a fraction of the nominal flow rate (e.g. Q means the nominal mass flow rate and 0.25Q means one fourth of the nominal flow rate).The resulting average oil velocity $-\bar{u}_G$—has been evaluated dividing the volumetric oil flow rate (at inlet temperature conditions) over the cross sectional area (of section G) through which the total mass flow is forced to flow through. The average oil velocity is a common winding design parameter in large shell-type transformers. The corresponding average Reynolds number, Re, has been evaluated using this average oil velocity (evaluated over section G) and the oil density at the inlet temperature of 60 °C.

The average Reynolds range observed is well within the laminar flow limit attributed to circular pipes ($Re < 2100$). It is noteworthy that, due to unequal flow distribution between the fluid channels, local Reynolds numbers three times higher than the average value have been observed, but even for simulation 2Q a maximum local Reynolds number of 562 was obtained, well within the laminar flow limit. At the inlet the pressure was observed to be approximately uniform different mass flow rates. Thus instead of finding the adequate total pressure value iteratively for each particular mass flow rate, a fixed dimensionless mass flow profile (obtained from one simulation) was used in the others with different target mass flow rate values.

Table 4.10 Boundary conditions, mesh and most relevant solver parameters

Inlet	Mass flow inlet	Variable. Target value with a fixed dimensionless profile
	Temperature	60 °C
Outlet	Static pressure	0 Pa
Hot-Plate	Constant heat flux	15.2 W.dm^{-2}
	Momentum	No-slip condition
Other Walls	Momentum	No-slip condition
	Heat transfer	Adiabatic
Symmetries	XY plane at middle height of the coil (Fig. 4.23a)	
	XZ plane along the pressboard washer (Fig. 4.23b)	
Mesh	Mesh6	
Solver	Pressure Based Solver	
	Coupled Pressure-Velocity^a (CFL = 200 and URF = 1.0)	
	2nd Order Discretization Schemes	
	Gradients with Least Square Cell Based	

^{a}CFL—Courant Flow Number; URF—Under-Relaxation Factors for Density, Body Forces and Energy

Table 4.11 Range of target mass flow rates imposed

Sim. ID	q_{inlet} [kg.s^{-1}]	$\bar{u}_G$ [m.s^{-1}]	Re
2Q	0.0408	0.187	187.4
1.38Q	0.0282	0.130	129.9
Q	0.0204	0.093	93.7
0.5Q	0.0102	0.046	46.5
0.25Q	0.0051	0.023	23.3

The mass flow profile in Fig. 4.29 shows the oil preferably flows closer to the curved bottom area (Fig. 4.23b), which corresponds to the higher value plotted in Fig. 4.29. This is compliant with previous findings from (Gomes et al. 2007a, b). The results from these CFD simulations are described in the next subsection.

4.3.1.4   Results

The CFD results herein shown refer to the maximum and minimum flow rates (2Q and 0.25Q, respectively). The results are reported in CFD maps and, for a better qualitative comparison between the different simulations the results are shown in terms of two variables: u is the oil velocity magnitude; and θ is a scalar representing the dimensionless oil temperature defined by

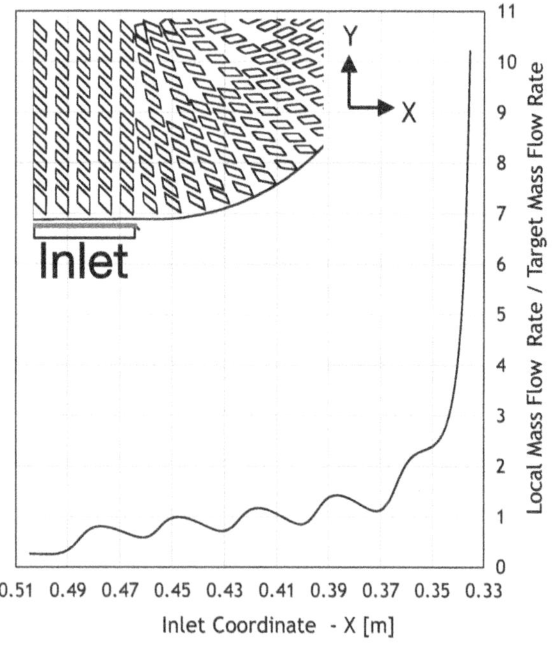

Fig. 4.29 Dimensionless flow profile imposed in the inlet surfaces. Originally extracted at middle height

$$\theta = \frac{(T_{oil,z,i} - T_{inlet})}{(T_{oil,2} - T_{inlet})} \tag{4.36}$$

where $T_{oil,z,i}$ is the oil temperature at the vertical position i, T_{inlet} is the oil temperature at the inlet and $T_{oil,2}$ is the mass weighted average oil temperature at the outlet.

All maps are shown within the same scale limits listed in Table 4.12.

Figure 4.30 shows velocity magnitude maps, in a plane located at the middle height between the hot plate and the opposite colder surface of the pressboard washer (XY plane with constant Z coordinate = 0.000975 m), for simulations at 0.25Q and 2Q.

The oil velocity maps from Fig. 4.30 show increased blue coloured regions for the highest flow rate, meaning that for 2Q the number of low velocity regions is higher compared with 0.25Q. These low velocity regions are preferably located in the wake areas of the spacers which correspond to the radial channels.

An additional useful scalar variable is the mass flow rate fraction, X_F, in each fluid channel, defined by the following expression:

$$X_F \, (\%) = \frac{q_{channel,i}}{q_{total,inlet}} \times 100 \tag{4.37}$$

where $q_{channel,i}$ is the mass flow rate entering a specific fluid channel, and $q_{total,inlet}$ is the total mass flow rate entering the domain. X_F is plotted in Fig. 4.31b for consecutive fluid channels belonging to the same row of spacers shown in Fig. 4.31a.

Identical mass flow distributions are observed for the minimum and mass flow rates simulated. This is in contrast with core-type windings, where the mass flow rate distribution among the channels of a zig-zag disc-type winding has been found to be significantly sensitive to the total mass flow rate entering the domain (Wu et al. 2012). This is an important observation revealing, that the local pressure losses do not affect significantly the mass flow distribution among the fluid channels, thus enabling simpler assumptions to describe the pressure drop between consecutive nodes of the fluid network. Currently in FluSHELL, the pressure drop between consecutive nodes is purely resistive, as referred in Eq. (4.7).

The local fluid flow patterns inside specific fluid channels are shown in Fig. 4.32. The 24-4 transverse and radial fluid channels, identified in Fig. 4.27, are used as an example.

Results in Fig. 4.32 illustrates for 2Q show that, a significant fraction of the fluid entering the transverse channel is displaced to the vicinity of one wall while in the radial channel the recirculation region is increased. For these reasons, even though

Table 4.12 Scale limits applied to the CFD maps

Field	Lower limit	Upper limit
Velocity magnitude (u)	0	$3 \times \bar{u}_G$
Dimensionless oil temperature (θ)	0	2

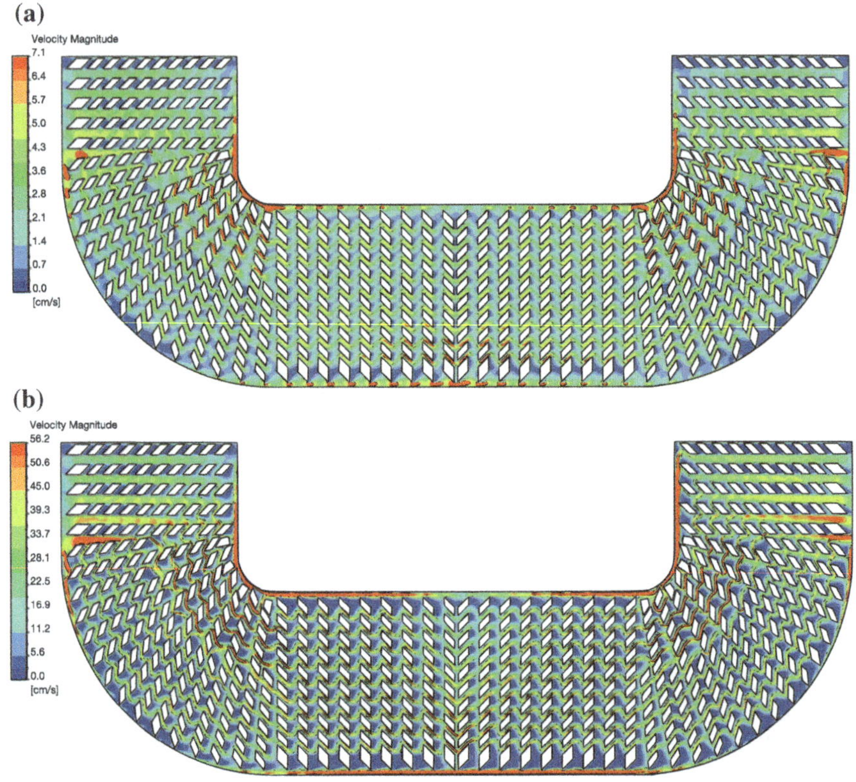

Fig. 4.30 Velocity Magnitude Maps for **a** 0.25Q and **b** 2Q in a plane located at middle height (Z = 0.000975 m)

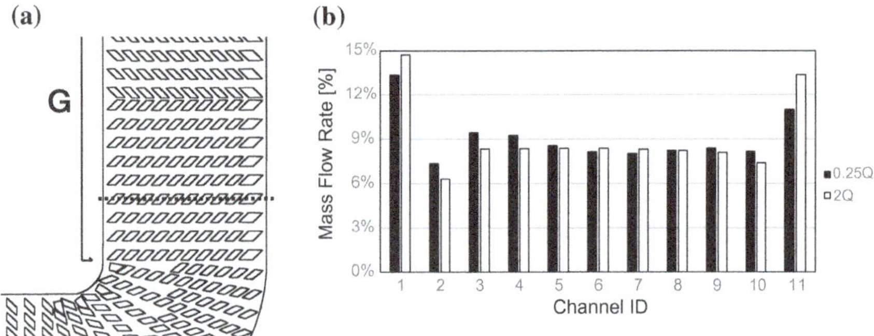

Fig. 4.31 **a** Consecutive fluid channels belonging to the same row of spacers and **b** corresponding mass flow rate distribution

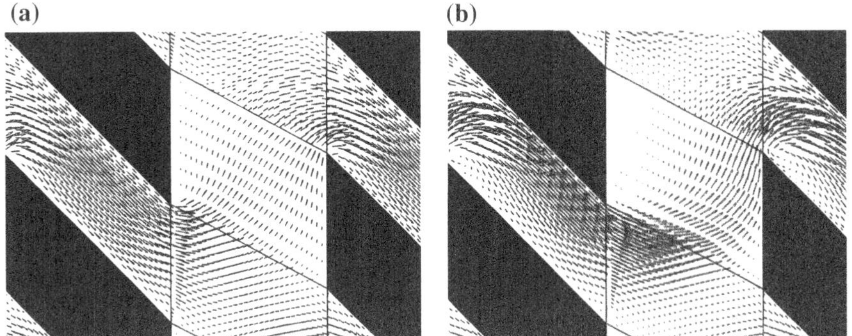

Fig. 4.32 Velocity magnitude vectors for **a** 0.25Q and **b** 2Q plotted in a plane located at middle height (Z = 0.000975 m)

2Q involves higher velocity magnitudes than 0.25Q, the onset of such local non-uniformities can lead to lower heat transfer efficiencies. In FluSHELL these local effects are lumped together in a friction coefficient representing the whole fluid channel.

Figure 4.33 describes the dimensionless oil temperature differences along the height of the transverse channel 24-4, from 0.25Q to 2Q. The values extracted correspond to the centroids of the 6 mesh volumes used along the height ($+Z$ Direction). The height is divided by 0.00195 m, thus represented in the plot as a dimensionless vertical position (e.g. 1.0 corresponds to the hot plate).

The dimensionless oil temperature depicted in Fig. 4.33b suggests that, depending on the vertical position, the conclusions can be misleading. For example, at the vertical position 0.6 the dimensionless oil temperature is identical for the

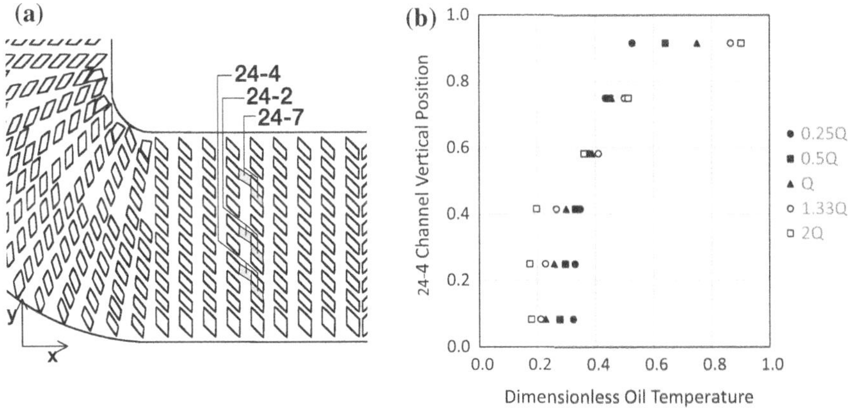

Fig. 4.33 Dimensionless Oil Temperature differences along the +Z Direction **a** location of 24-4 fluid channel **b** values plotted for transverse fluid channel 24-4

whole range of flow rates. However, at the vertical position 0.92 the dimensionless oil temperature is 0.53 for 0.25Q and 0.90 for 2Q. In this latter case, for 2Q, the dimensionless temperature reaches a maximum around 4.0 meaning the oil at that location is hotter than in the outlet (refer to the definition in Eq. (4.36)). Hence, Fig. 4.33b suggests the dimensionless oil temperature might be an adequate oil mixing indicator if analysed at a proper vertical position—corresponding to the nearest hot plate centroid.

Figure 4.34 shows dimensionless oil temperature maps, for 0.25Q and 2Q, in a plane coincident with the nearest hot plate centroids.

Figure 4.34 shows that the previous increased blue coloured regions for 2Q (in Fig. 4.30b) coincide with increased red coloured regions in Fig. 4.34b corresponding to higher dimensionless oil temperatures. Thus, even though the temperature magnitudes are lower, for 2Q a less homogeneous oil temperature distribution is observed.

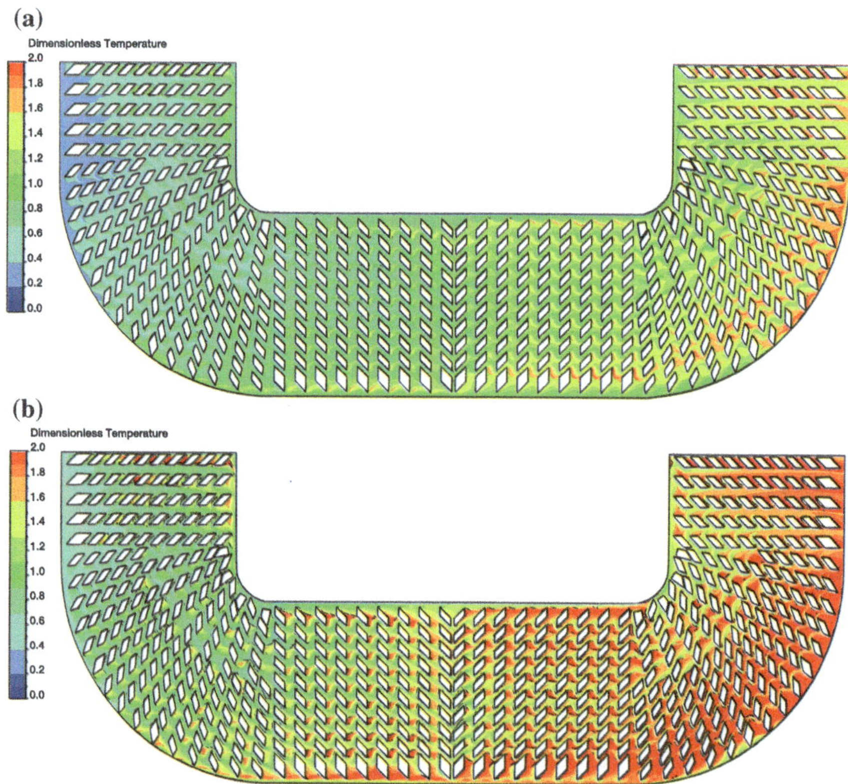

Fig. 4.34 Temperature maps for **a** 0.25Q and **b** 2Q in a plane located at Z = 0.001787 m

In FluSHELL these local effects are lumped together in a heat transfer coefficient representing the whole fluid channel.

4.3.2 Determination of Correlations

This section describes the methodology of extracting the corresponding friction and heat coefficients using the data from the CFD simulations.

The process of extracting transport properties such as friction and heat transfer coefficients involve computing a significant number of variables over many surfaces. Thus, some sample fluid channels have been chosen according to the locations depicted in Fig. 4.35 and coloured in blue.

A total of 308 sample fluid channels are distributed over the whole domain in symmetric locations: 168 radial channels and 140 transverse channels. Figure 4.36 identifies the quantities that have been extracted from each sample fluid channel in order to further compute friction and heat transfer coefficients.

For this purpose, several surfaces have been created in the sample fluid channels, during the pre-processing stage. These surfaces are identified in Fig. 4.37 for the transverse channel 24-4.

The shear stress magnitude is computed at the top, bottom and side walls identified in Fig. 4.37. The top wall corresponds to the *hot plate* depicted in Fig. 4.26 where a constant heat flux has been imposed, thus being the single heated wall. The area averaged temperature of this wall is also computed. Additional surfaces have been created at the inlet and outlet of the fluid channels in order to extract mass flow rates and oil temperatures.

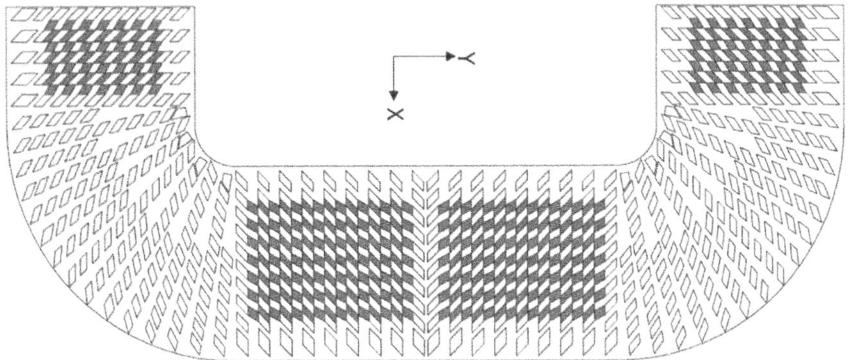

Fig. 4.35 Sample fluid channels coloured in blue

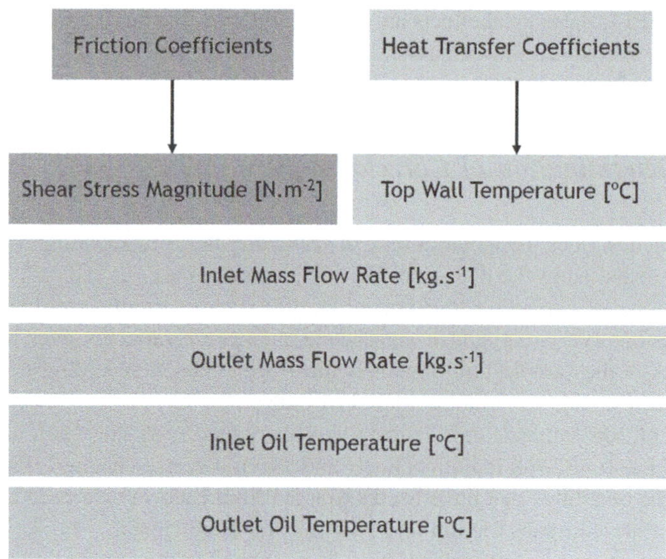

Fig. 4.36 Diagram of the variables extracted from the CFD simulations

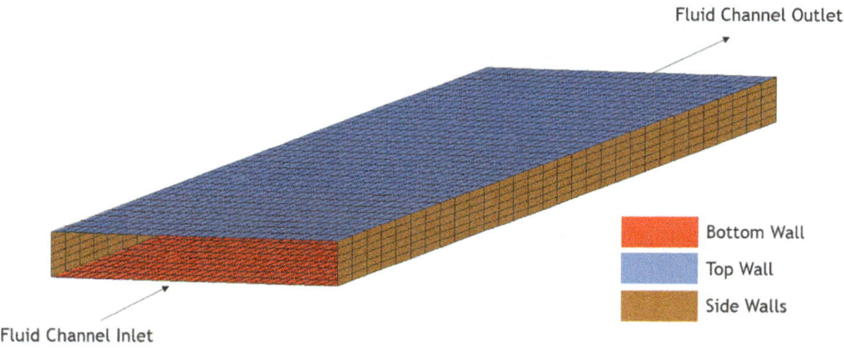

Fig. 4.37 Identification of the individual walls of each fluid channel used to extract data

4.3.2.1   Friction Coefficients

In fully developed laminar flow of an incompressible fluid, which is the case, the pressure drop inside a fluid channel equals the friction forces on its surrounding walls according to

$$A_f \Delta P = \sum_w \tau_w A_w \tag{4.38}$$

where A_f represents the cross sectional area of each fluid channel, A_w is the area of the walls where friction forces are acting, ΔP is the fluid pressure drop, τ_w is the shear stress magnitude on the wetted walls. On the other hand, the fluid pressure drop due to friction can be described by

$$\Delta P = \frac{1}{2} f \rho \frac{L_{ch}}{d_h} u_{ch}^2 \tag{4.39}$$

where f is the friction coefficient, L_{ch} is the characteristic hydraulic length of the fluid channel, d_h is the hydraulic diameter of the fluid channel and u_{ch} is the average oil velocity in the channel.

Combining Eqs. (4.38) and (4.39), the friction coefficient can be expressed as

$$f = \frac{2d_h}{u^2 L_{ch} \rho} \frac{\displaystyle\sum_w \tau_w A_w}{A_{f,ch}} \tag{4.40}$$

where τ_w is the shear stress magnitude obtained from CFD and computed in the walls identified in Fig. 4.37 (top, bottom and side walls); $A_{f,ch}$ is the flow area, L_{ch} is the characteristic hydraulic length, d_h is the hydraulic diameter of the fluid channels listed in Table 4.2 and u_{ch} is the average fluid velocity obtained from CFD computed as the mass flow rate in the channel divided by the flow area. The ρ represents the fluid density evaluated for the average fluid temperature in the channel computed by CFD.

During this process, the mass flow rate comparison between the inlet and outlet surfaces led to an apparent imbalance in 13% of the sample fluid channels. The imbalance is said to be apparent because the simulations show negligible local residuals for each equation solved. Figure 4.38 shows the velocity magnitude vectors at the inlet and outlet surfaces of a particular transverse channel for the 0.25Q simulation.

Reversed velocity vectors are observed in both surfaces due to a recirculating fluid flow pattern. As a result, a recirculation factor has been defined according to the following equation

$$RF_{ch} = \frac{\left| \int \rho_{ch,in} \overrightarrow{u_{ch,in}} \cdot dA - \int \rho_{ch,out} \overrightarrow{u_{ch,out}} \cdot dA \right|}{\rho_{ch} u_{ch} A_{f,ch}} \Leftrightarrow RF_{ch} = \frac{\left| q_{ch,in} - q_{ch,out} \right|}{q_{ch}} \tag{4.41}$$

The recirculation factor, RF_{ch}, is a quality parameter of the fluid channel data. Whenever a fluid channel exhibits a recirculation factor higher than 0.02 (2%), the friction and heat transfer results from that fluid channel are discarded. This recirculation would unnecessarily increase the dispersion of the data sampled and decrease the quality of further correlations.

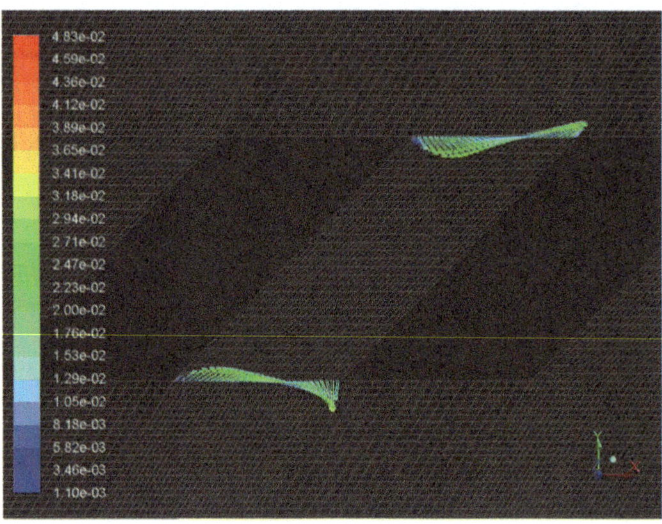

Fig. 4.38 Velocity magnitude vectors at the inlet and outlet surfaces of transverse channel 24-4 for the 0.25Q simulation

The friction coefficients for transverse and radial channels (the sample fluid channels indicated in Fig. 4.35), computed according to Eq. (4.40) for simulations 0.25Q up to 2Q are plotted in Fig. 4.39.

A recirculation factor higher than 2% has been observed in 13% of the channels, hence the data from these channels has been discarded. Finally, an average number of 268 sample fluid channels has been used from each of the 5 CFD simulations (0.25Q, 0.5Q, 1Q, 1.38Q and 2Q). A total number of valid 1380 data points have been used.

The friction coefficients extracted from CFD have been further correlated using the least squares method. The generic fitting expression used is

$$f = \frac{C}{Re} \tag{4.42}$$

where C is the single fitting parameter.

The best fitting expression for the transverse channels and the radial channels are given by

$$f_{T,CFD} = \frac{68.4}{Re} \tag{4.43}$$

$$f_{R,CFD} = \frac{75.2}{Re} \tag{4.44}$$

The analytical solution for fully developed flow between infinite parallel plates is

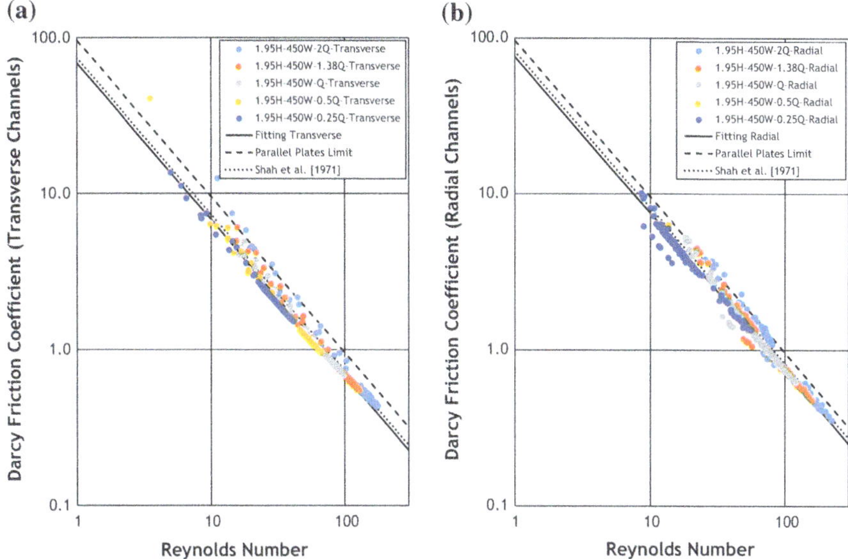

Fig. 4.39 Friction coefficients extracted from CFD for: **a** transverse channels and **b** for radial channels

$$f_{Plates} = \frac{96}{Re} \tag{4.45}$$

which is also plotted in Fig. 4.39 as well as the analytical solution for rectangular ducts with similar aspect ratios. This solution can be found in a comprehensive work from (Shah and London 1971) for different compact heat exchangers configurations.

For the aspect ratio of the transverse channels 4.24, this analytical solution is given by

$$f_{4.24,Shah} = \frac{73.8}{Re} \tag{4.46}$$

For the aspect ratio of the radial channels 7.52, this analytical solution is given by

$$f_{7.52,Shah} = \frac{81.6}{Re} \tag{4.47}$$

The analytical solutions from (Shah and London 1971) assume fully developed flow. Both plots in Fig. 4.39 show that the CFD results and these latter analytical solutions are identical, therefore reinforcing that the flow is hydraulically fully developed in most the channels (as observed in Table 4.4).

4.3.2.2   Heat Transfer Coefficients

The heat transferred from a hot wall to a flowing fluid can be expressed as

$$Q = UA(T_{hotwall} - T_{fluid}) \Leftrightarrow \frac{Q}{A} = U(T_{hotwall} - T_{fluid}) \qquad (4.48)$$

where $\frac{Q}{A}$ is the heat flux, U is the heat transfer coefficient, A is the effective heat transfer area, $T_{hotwall}$ is the average hot wall temperature and T_{fluid} is the fluid bulk temperature.

Once the heat flux and the temperatures are known, the heat transfer coefficients can be computed from CFD using

$$U = \frac{Q}{A(T_{hotplate} - T_{ch,inlet})} \qquad (4.49)$$

where $\frac{Q}{A}$ is the heat flux imposed as a boundary condition in each CFD simulation (constant value of 15.2 W.dm^{-2} for all simulations—Table 4.10), $T_{hotplate}$ is the area weighted average temperature extracted from the top wall of each fluid channel and $T_{ch,inlet}$ is the area weighted average temperature extracted at the inlet of each fluid channel (refer to the control surfaces of each fluid channel identified in Fig. 4.37).

The heat transfer coefficients are then expressed in terms of the dimensionless Nusselt Number according to

$$Nu = \frac{Ud_h}{k} \qquad (4.50)$$

where k is the thermal conductivity of the fluid evaluated at the inlet temperature of each channel, is d_h the hydraulic diameter of each channel (refer to definition in Table 4.4) and U is the heat transfer coefficient computed using Eq. (4.49).

The heat transfer coefficients, for transverse and radial channels, expressed in terms of Nusselt Numbers, computed according to Eq. (4.50) for simulations 0.25Q up to 2Q, are plotted in Fig. 4.40.

The Nusselt Numbers shown in Fig. 4.40 have been computed in the same sample fluid channels, respecting the same recirculation criteria, as the friction coefficients. Finally, an average number of 268 sample fluid channels has been used from each of the 5 CFD simulations (0.25Q, 0.5Q, 1Q, 1.38Q and 2Q). Which amounts to the same total number of valid 1380 data points.

The Nusselt numbers extracted from CFD have been further correlated using the least squares method. The generic fitting expression used assumes the form

$$Nu = ARe^{n} + Nu_{lim} \qquad (4.51)$$

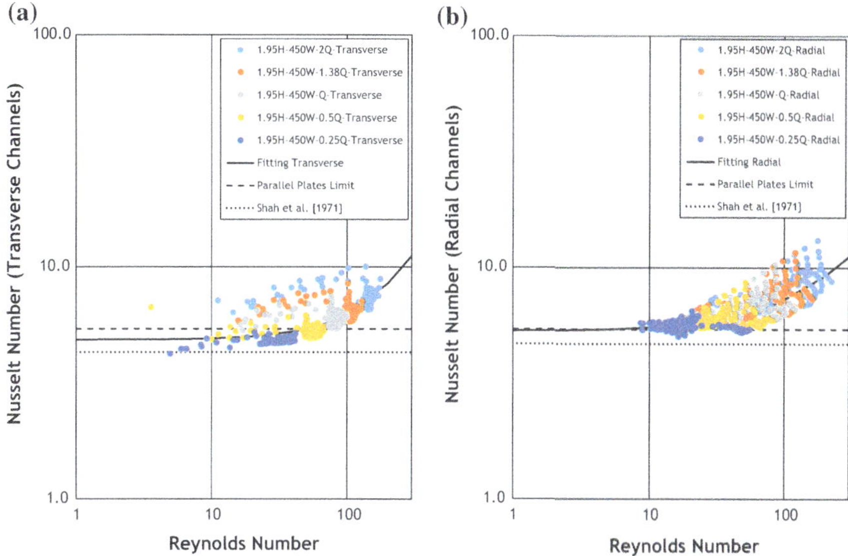

Fig. 4.40 Nusselt Numbers extracted from CFD for: **a** transverse channels and **b** for radial channels

where A, n and Nu_{lim} represent the fitting parameters. Nu_{lim} also represents the asymptotic Nusselt number for fully developed temperature profile.

The best fitting expression for the transverse channels and the radial channels are given by

$$Nu_T = 0.00275Re^{1.36} + 4.84 \tag{4.52}$$

$$Nu_R = 0.0216Re^{0.982} + 5.30 \tag{4.53}$$

The analytical solution for fully developed temperature profiles between constant heat flux infinite parallel plates

$$Nu_{Plates} = 5.39 \tag{4.54}$$

is also plotted in Fig. 4.40 as well as the analytical solution for rectangular ducts with similar aspect ratios. This solution can be found in the same comprehensive work from (Shah and London 1971) for different compact heat exchangers configurations. For the aspect ratio of the transverse channels 4.24 the analytical solution is

$$Nu_{4.24,Shah} = 4.28 \tag{4.55}$$

For the aspect ratio of the radial channels 7.52, the analytical solution is

$$Nu_{7.52,Shah} = 4.66 \qquad\qquad (4.56)$$

These analytical solutions from (Shah and London 1971) assume fully developed temperature profiles, meaning the Nusselt assumes a constant value and have been computed assuming a single sided constant heat flux wall, which is also the case for the CFD simulations conducted. Both plots in Fig. 4.40 show that the CFD results tend asymptotically to values which are close to the analytical solutions. It is also noticeable that the CFD results are not constant, reinforcing the temperature profiles in this geometry are not fully developed.

The friction coefficients expressed in Eqs. (4.43) and (4.44) as well as the Nusselt numbers expressed in Eqs. (4.52) and (4.53) have been implemented in the FluSHELL tool.

4.4   FluSHELL Results

This section describes the FluSHELL tool developed and intends to show the main outputs of the tool further validated in Chap. 5.

The FluSHELL tool has been developed using VBA for Microsoft Excel®. The main worksheet is shown in Fig. 4.41.

This worksheet is organized in 3 main areas:

- *Simular* which permits starting a new simulation of a coil/washer system or configuring a pre-existing one;
- *Visualizar Rede* which, after some calculation steps, permits a graphical visualization of the fluid and solid networks;
- *Analisar Resultados* which permits plotting the temperature distribution both in the fluid and in the coil.

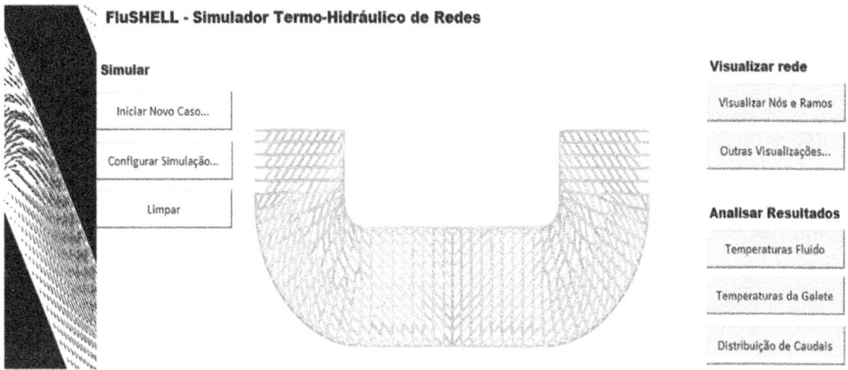

Fig. 4.41 Main Excel worksheet—main interface of the FluSHELL tool

Figure 4.42 shows the input form to start inputting data after initiating a new simulation.

The information needed to initiate a simulation is depicted in Fig. 4.4. The first step requires importing data for each spacer which includes the coordinates of its centroid, type, rotation and size. This information is inputted through a text file.

The second step requires inputting data to define the coil, the washer and the insulation frames as shown in Fig. 4.43.

Four previously disabled regions are now enabled in the main input form:

- *Composição do Feixe*, where the turn is defined. For example, the number copper conductors it comprises, its arrangement, its dimensions, materials…
- *Galete*, where the coil is defined. For example, the number of turns it comprises and other global geometric dimensions.
- *Rodela*, where the thickness of the spacers and the number of fluid inlets is defined.
- *Moldados*, where the type of insulation frame is defined, both the internal and external frames. In this region the type of frame is defined together with its dimensions. The inner and outer insulation frames might be different from each other.

The next step is the generation of both fluid and solid networks (Fig. 4.44).

After generating the networks and before starting the simulation it is possible to plot the geometry (washers with spacers and insulation frames), the fluid nodes, the

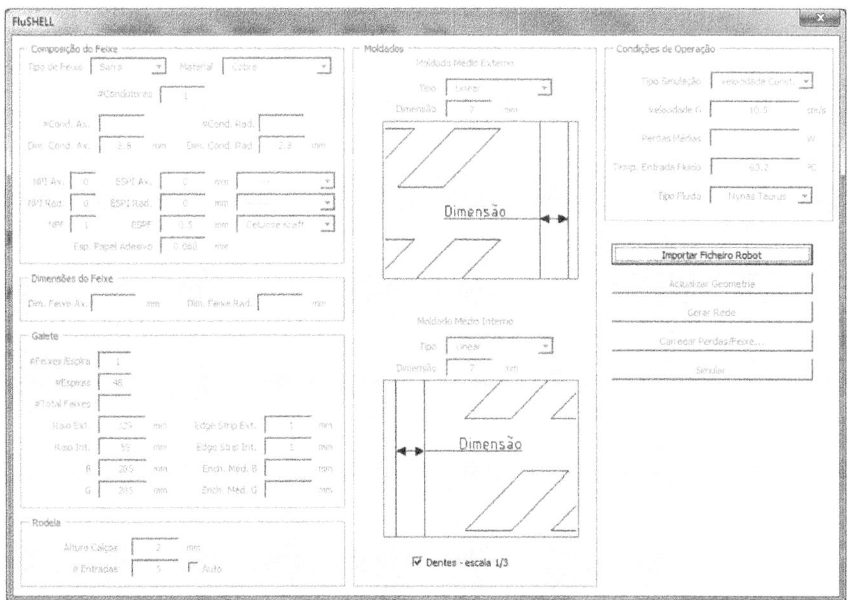

Fig. 4.42 Initial form to input data. Importing the spacers text file

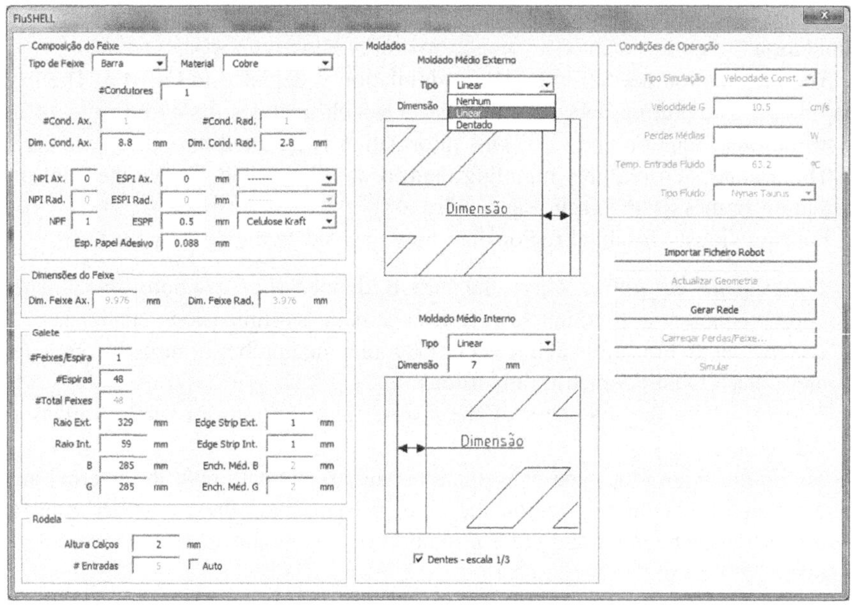

Fig. 4.43 Initial form to input data. Defining turns, coil, washer and insulation frames

fluid branches, the fluid channels, the turns and the connections between neighbouring turns. Figure 4.45a shows the plot of fluid nodes and branches numbered, Fig. 4.45b shows another plot with the fluid channels and Fig. 4.45c shows the plot of the turns where it is possible to identify the different types of turn segments—the orange turn segments contacting with radial fluid channels, the green turn segments contacting with transversal and the grey contacting with either spacers or insulation frames.

This provide means for a first assessment about whether the interpretation of the coil/washer system is correct or not. Then the final step involves defining the operating conditions (Fig. 4.46).

The operating conditions include (Fig. 4.47):

- the type of simulation to be conducted: constant velocity or constant pressure. The velocity solver requires an additional definition of the target average fluid velocity (evaluated over G section of the coil) and the pressure solver requires defining the total pressure at the inlet of the system.
- the fluid to be used as well as its temperature at the inlet.
- the heat losses: uniformly distributed or not. In the case of non-uniform losses it is required to import the heat loss distribution per turn using an external text file.

Then the simulation can be started. Figure 4.48 shows the global results for EXP1 conditions. The coil/washer system simulated represents the scale model.

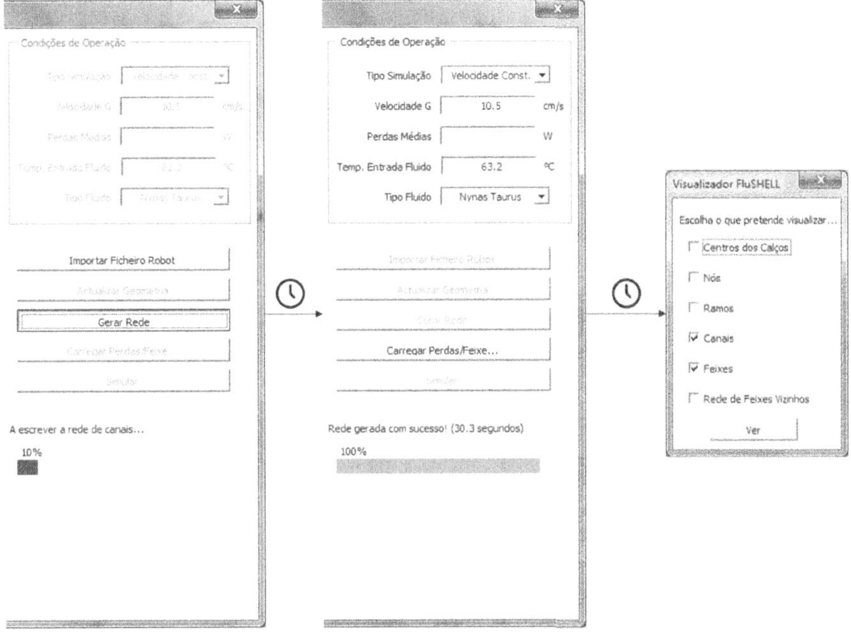

Fig. 4.44   Generation of the fluid and solid networks. Visualization of both networks

The global results are separated in 7 regions:

- *Ponto de Operação* with information about the volumetric flow rate at the inlet and the corresponding pressure.
- *Temperaturas Fluido* where the minimum, average and maximum temperatures in fluid are reported.
- *Temperaturas Galete* where the minimum, average and maximum temperatures in the coil are reported.
- *Gradientes* where the average and maximum thermal gradients can be found. The corresponding hotspot factor is also shown.
- *Informação Solver* with information about the performance of the solver (calculation time, total time elapsed, number of global iterations, total number of thermal iterations, …).
- *Resíduos Simulação* where the residuals of the mass and energy balances are reported. These are quality indicators about the physical coherence of each simulation.
- *Informação Complementar* with complementar information about the maximum Reynolds and maximum oil velocity observed inside the coil.

Besides this global results, FluSHELL computes the local temperature results for each turn, as shown in Fig. 4.48. The scale model coil is composed of 48 turns.

The local results include the average and maximum temperatures for each turn.

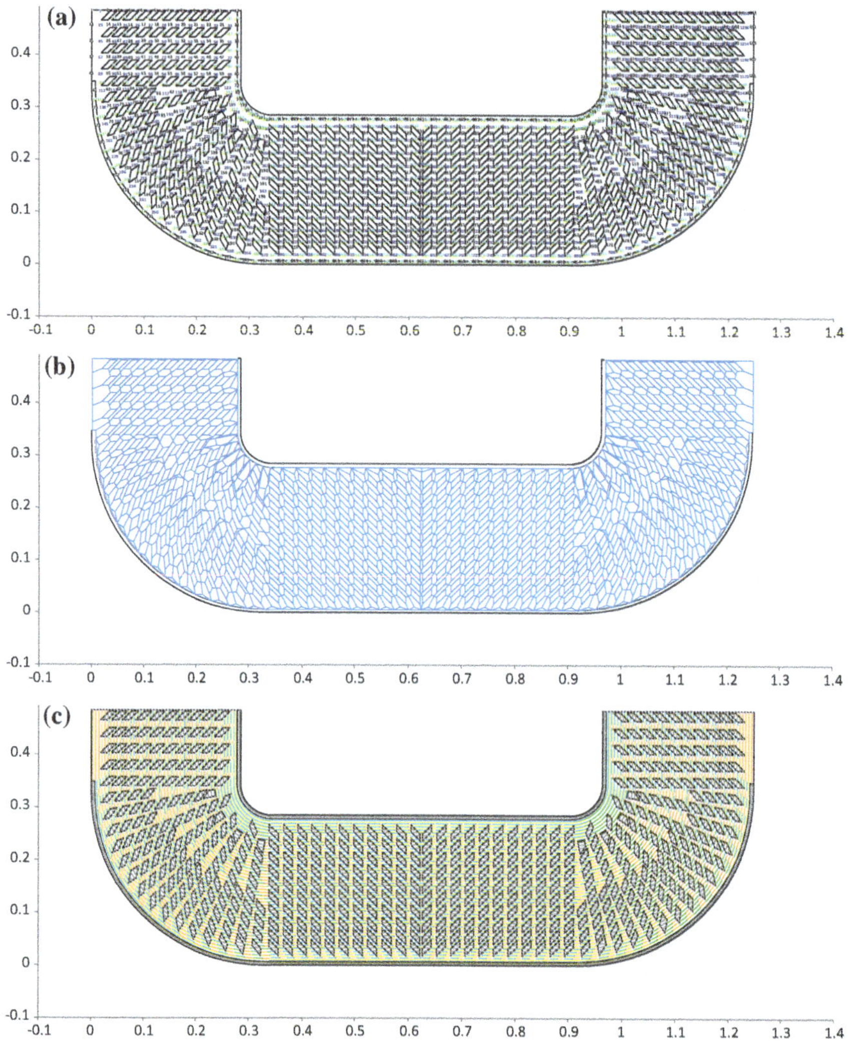

Fig. 4.45 FluSHELL plots: **a** numbered nodes and branches; **b** fluid channels and **c** turns

Finally, it is also possible to plot the temperature distribution in the coil, the temperature distribution in the fluid and the mass flow fraction distribution. Figure 4.49a shows a plot of the temperature distribution in the coil and Fig. 4.49b shows a plot of the mass flow fractions, X_F.

The temperature contours provide means for analyzing and identifying the location of the hottest temperature occurring in the coil.

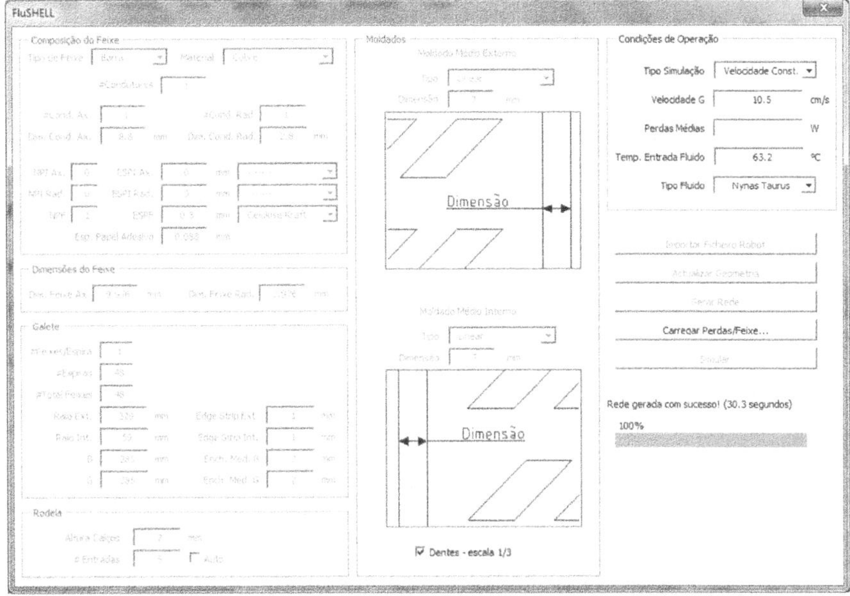

Fig. 4.46 Initial form to input data. Setting the operating conditions

RESULTADOS GLOBAIS:

PONTO DE OPERAÇÃO:			INFORMAÇÃO SOLVER:		
Caudal Vol. Entrada	0.42	m³/ h	Tempo Cálculo	2.8	minutes
Pressão Entrada	0.54	o.c.m.	Tempo Total	3.0	minutes
	4500	Pa	No. Iter. Global	12	
			No. Iter. Term. Total	1920	
Velocidade Méd. G	10.5	cm/s	No. Iter. Term. Máx.	500	
Potência/Área Molhada	25.6	W/dm²	Nr. Canais LowFlow	0	(em 1268)
Factor Área Molhada	68	%			

			RESÍDUOS SIMULAÇÃO:		
			Caudal Mássico Entrada Sim.	2.49E-02	kg/s
TEMPERATURAS FLUIDO:			Caudal Mássico Saída Sim.	2.49E-02	kg/s
Temp. Entrada Fluido	63.2	°C	Massa	8E-12	%
Temp. Méd. Fluido	70.5	°C	Potência Sim.	737	W
Temp. Saída Fluido	77.9	°C	Energia	2E-02	%

TEMPERATURAS GALETE:			INFORMAÇÃO COMPLEMENTAR:		
Temp. Min. Galete	77.7	°C	Reynolds G	83	
Temp. Méd. Galete	90.6	°C	Reynolds Máx.	693	
Temp. Máx. Galete	123.3	°C	Velocidade Máx.	122.0	cm/s

GRADIENTES:		
Gradiente Méd. Galete	20.1	°C
Gradiente Máx. Galete	45.4	°C
Hot Spot Factor	2.26	

Fig. 4.47 FluSHELL global results

| DADOS DE ENTRADA: | | | | | | | RESULTADOS LOCAIS: | | | |
# Galete	# Feixe	RI2[W]	PT[W]	PT[W/M3]	Q factor	Lfeixe (m)	Temp. Méd. Fx. (°C)	Temp. Máx. Fx. (°C)	Grad. Méd. Fx. (°C)	Grad. Máx. Fx. (°C)
1	1		43.0	842050	1.00	2.1	114.9	118.2	44.3	40.3
1	2		43.8	842050	1.00	2.1	87.2	92.2	16.6	14.3
1	3		44.6	842050	1.00	2.1	83.3	88.0	12.7	10.1
1	4		45.3	842050	1.00	2.2	84.7	88.6	14.1	10.8
1	5		46.1	842050	1.00	2.2	87.7	91.6	17.1	13.7
1	6		46.9	842050	1.00	2.3	89.9	94.0	19.3	16.1
1	7		47.7	842050	1.00	2.3	88.9	93.4	18.4	15.5
1	8		48.5	842050	1.00	2.3	87.3	91.4	16.8	13.5
1	9		49.3	842050	1.00	2.4	88.9	93.3	18.3	15.4
1	10		50.0	842050	1.00	2.4	91.1	94.9	20.5	17.0
1	11		50.8	842050	1.00	2.4	90.5	94.8	20.0	16.9
1	12		51.6	842050	1.00	2.5	88.3	93.1	17.8	15.2
1	13		52.4	842050	1.00	2.5	88.2	92.7	17.7	14.8
1	14		53.2	842050	1.00	2.6	90.0	94.1	19.5	16.2
1	15		54.0	842050	1.00	2.6	89.9	95.0	19.3	17.1
1	16		54.8	842050	1.00	2.6	87.8	92.9	17.3	15.0
1	17		55.5	842050	1.00	2.7	87.5	92.0	17.0	14.2
1	18		56.3	842050	1.00	2.7	89.3	93.8	18.7	15.9
1	19		57.1	842050	1.00	2.8	90.0	95.6	19.4	17.7
1	20		57.9	842050	1.00	2.8	88.2	93.6	17.6	15.7
1	21		58.7	842050	1.00	2.8	87.5	92.2	16.9	14.3
1	22		59.5	842050	1.00	2.9	88.9	93.6	18.3	15.7
1	23		60.3	842050	1.00	2.9	90.0	95.6	19.5	17.7
1	24		61.0	842050	1.00	2.9	89.0	94.5	18.4	16.6
1	25		61.8	842050	1.00	3.0	88.7	94.8	18.1	16.9
1	26		62.6	842050	1.00	3.0	90.4	97.3	19.8	19.4
1	27		63.4	842050	1.00	3.1	91.6	97.8	21.1	19.9
1	28		64.2	842050	1.00	3.1	90.3	95.7	19.8	17.8
1	29		65.0	842050	1.00	3.1	89.4	95.8	18.9	17.9
1	30		65.7	842050	1.00	3.2	91.0	98.3	20.4	20.4
1	31		66.5	842050	1.00	3.2	92.5	98.8	21.9	20.9
1	32		67.3	842050	1.00	3.2	91.1	96.7	20.6	18.8
1	33		68.1	842050	1.00	3.3	89.5	95.8	19.0	18.0
1	34		68.9	842050	1.00	3.3	90.5	97.9	20.0	20.0
1	35		69.7	842050	1.00	3.4	92.1	98.7	21.6	20.8
1	36		70.5	842050	1.00	3.4	91.5	97.4	20.9	19.5
1	37		71.2	842050	1.00	3.4	90.2	95.8	19.6	17.9
1	38		72.0	842050	1.00	3.5	91.1	97.6	20.5	19.7
1	39		72.8	842050	1.00	3.5	92.5	98.7	21.9	20.8
1	40		73.6	842050	1.00	3.5	91.2	97.2	20.6	19.3
1	41		74.4	842050	1.00	3.6	90.2	96.2	19.7	18.3
1	42		75.2	842050	1.00	3.6	91.2	97.7	20.6	19.8
1	43		76.0	842050	1.00	3.7	92.2	98.7	21.7	20.8
1	44		76.7	842050	1.00	3.7	90.1	96.8	19.6	18.9
1	45		77.5	842050	1.00	3.7	87.4	94.3	16.9	16.4
1	46		78.3	842050	1.00	3.8	86.8	93.8	16.2	15.9
1	47		79.1	842050	1.00	3.8	90.2	97.5	19.7	19.6
1	48		79.9	842050	1.00	3.8	112.2	123.3	41.7	45.4

Fig. 4.48 FluSHELL local results

4.5   Conclusions

This chapter describes the development of the first known thermal-hydraulic network tool for shell-type transformers, the FluSHELL tool. The tool has been developed to solve a single coil/washer system where both cooling sides of the coil are assumed independent and symmetrical. Unlike CFD, this tool describes the coil/washer system through simpler algebraic equation systems that use analogies with electrical circuits to compute pressures and temperatures. In that way, the tool is expected to show accurate predictions and be considerably faster than CFD.

The tool has been architected in the following way:

- First a topological model discretizes all the fluid and solid domains of the coil/washer system. The fluid domains are subdivided into several interconnected fluid channels which together form a fluid network of nodes and branches. In turn, the solid domains of the coil are subdivided into several interconnected turn segments which together form a solid network of nodes and branches. The

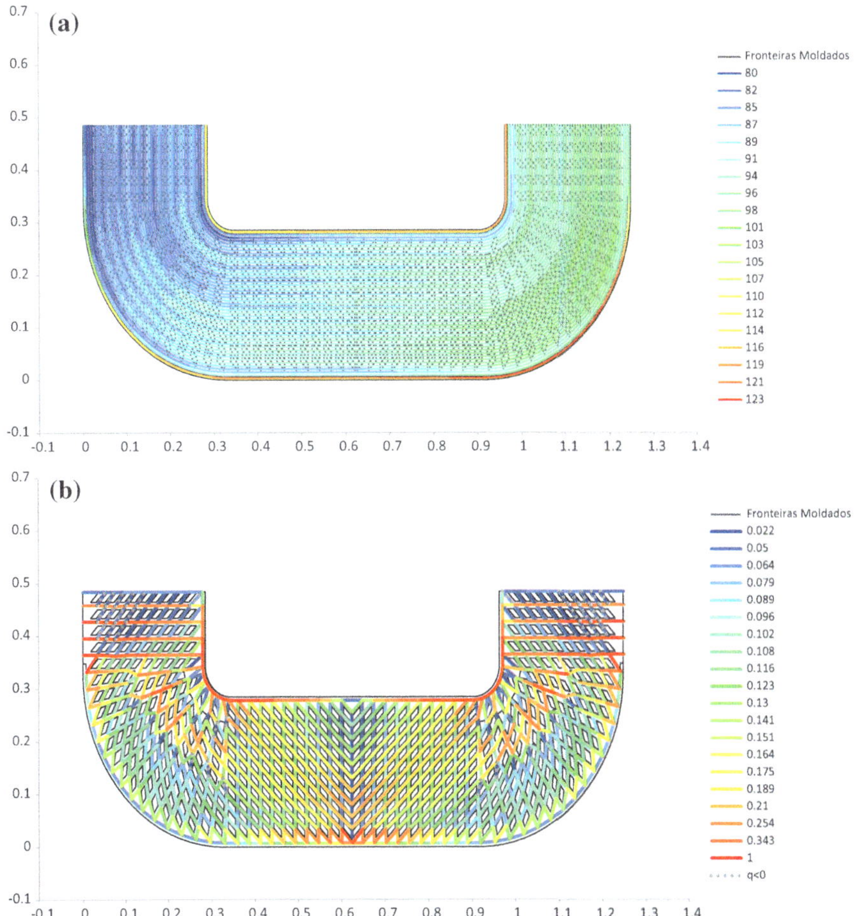

Fig. 4.49 FluSHELL plots: **a** coil temperatures and **b** mass flow rate fractions

solid network is generated through geometrical intersections between the coil and the fluid channels (Coil-Fluid interfaces), between the coil and the spacers as well as the insulation frames (Coil-Solid interfaces) and between neighbouring turns within the same coil (Coil-Coil interfaces). These interfaces enable establishing degrees of freedom for the heat transfer along all the relevant Cartesian coordinates.

- Then a hydrodynamic model uses the fluid network generated to implement an equation system describing the relationship between node pressures and branch flowrates. Now, this relationship considers exclusive resistive effects being the final hydraulic resistance not dependent on the mass flow rate. This an intrinsic characteristic of this transformer technology, which is supported in this chapter by simulations that show an identical mass flow distribution for highly different mass flow rates. The result of this model is the fluid pressure in each node of the fluid network and hence the respective mass flow rate in each branch.

- Afterwards a heat transfer model uses the solid network generated to implement three equation systems that describe the energy conservation in the fluid nodes, in the fluid branches and in the turn segments of the coil. Two fundamental heat transfer mechanisms have been considered in those equations: thermal conduction and thermal convection. Thus, the result of this model is the fluid temperature in each node and each branch as well as the solid temperature in each turn segment.

As the fluid and solid networks are interdependent, the resulting systems of equations are solved iteratively until predefined criteria of convergence are observed. The hydrodynamic model controls the global convergence of the simulation depending whether the pressure distribution in the fluid nodes is below or not the convergence criterion assumed.

Unlike CFD, the hydrodynamic and the heat transfer models of the FluSHELL tool do not model directly the fluid velocity and temperature profiles inside each element of the fluid network. Instead, the tool models the velocity and temperature variations between the hot wall and the bulk of each fluid channel using lumped transport properties such as friction and heat transfer coefficients. In this chapter these coefficients have been extracted from sets of parametric CFD simulations and its values have been compared against the available values in literature.

From this analysis, it has been observed that:

- There are two types of channels with distinct average aspect ratios and subsequently different friction coefficients: transverse and radial fluid channels. Both types of fluid channels show a good agreement between the friction coefficients extracted from CFD and the values available in literature for fully developed flows in rectangular shaped channels with the same aspect ratios. This observation reinforces the approximation of fully developed flow conditions inside each fluid channel of the network.
- The same distinction applies to the heat transfer coefficients, where the same segregation between transverse and radial channels led to different heat transfer coefficients. Due to the high Prandtl number of the mineral oil used as the cooling fluid, the Nusselt numbers in both types of channels have been observed to vary for different Reynolds numbers, indicating the fluid flow is thermally developing. However, the asymptotic Nusselt numbers for each type of channels are also in good agreement with the values available in literature for fully developed flows in rectangular shaped channels with the same aspect ratios.

The friction and heat transfer coefficients extracted from CFD have been correlated using fitting expressions. The expressions for transverse channels have been extracted in a Reynolds range between 4 and 176 while for radial channels the range is between 9 and 223. However, these expressions are expected to be valid for the whole laminar range and have been finally implemented in the FluSHELL tool. Thus, at this moment, FluSHELL tool exclusively distinguishes between transverse and radial channels independently of the local aspect ratio of each channel.

Globally, the FluSHELL tool developed can adapt the fluid network to different number of inlets, to different shapes/sizes of the insulation frames and to different spacers (number, position and size). The tool is also able to adapt the solid network to different dimensions and numbers of turns.

These characteristics are expected to cover most of the current designs used in commercial shell-type transformers, being its results validated against both experiments and CFD in Chap. 5.

References

Blume, L. F., Boyajian, A., Camilli, G., Lennox, T. C., Minneci, S., & Montsinger, V. M. (1951). *Transformer engineering: A Treatise on the theory, operation, and application of transformers*.

Campelo, H. M. R., Lopez-Fernandez, X. M., Picher, P., & Torriano, F. (2013). Advanced thermal modelling techniques in power transformers. Review and case studies. In *Advanced Research Workshop on Transformers*, Baiona (pp. 1–17).

Campelo, H. M. R., Baltazar, J. P. B., Oliveira, R. T., Fonte, C. M., Dias, M. M., & Lopes, J. C. B. (2015b). Extracting relevant transport properties using 3D CFD simulations of shell-type electrical transformers. In *International Symposium on Advances in Computational Heat Transfer*. New Jersey, USA: Begell House.

Campelo, H. M. R., Fonte, C. M., Lopes, R. C. L., Dias, M. M., & Lopes, J. C. B. (2012). Network modelling applied to CORE power transformers and validation with CFD simulations. In *International Colloquium Transformer Research and Asset Management*, Dubrovnik (pp. 1–16).

Campelo, H. M. R., Fonte, C. M., Sousa, R. G., Lopes, J. C. B., Lopes, R.C., Ramos, J., et al. (2009). Detailed CFD analysis of ODAF power transformer. In *International Colloquium Transformer Research and Asset Management*, Cavtat, Croatia (pp. 1–10).

Campelo, H. M. R., Oliveira, R. T., Fonte, C. M., Dias, M. M., & Lopes, J. C. B. (2014b). Modelling the hydrodynamics of cooling channels inside shell-type power transformers with CFD. In *International Colloquium Transformer Research and Asset Management*, Split (pp. 1–16).

Cigre. (2016). WG A2.38 Brochure, Transformers Thermal Modelling. Draft Version 5.

Coddé, J., van der Veken, W., & Baelmans, M. (2015). Assessment of a hydraulic network model for zig–zag cooled power transformer windings. *Applied Thermal Engineering, 80*, 220–228. https://doi.org/10.1016/j.applthermaleng.2015.01.063.

Declercq, J., & van der Veken, W. (1999). Accurate hot spot modeling in a power transformer leading to improved design and performance. *In IEEE Transmission and Distribution Conference* (pp. 920–924). New Orleans, USA: IEEE. https://doi.org/10.1109/TDC.1999.756172.

Del Vecchio, R. M., Poulin, B., Fegahli, P. T., Shah, D. M., & Ahuja, R. (2001). *Transformer design principles: With applications to core-form power transformers*. Gordon and Breah Science.

Desoer, C. A., & Kuh, E. S. (1969). *Basic circuit theory*.

Dias, M. M., & Payatakes, A. C. (1986). Network models for two-phase flow in porous media. Part I. Immiscible microdisplacement of non-wetting fluids. *Journal of Fluid Mechanics, 164*, 305–336. https://doi.org/10.1017/S0022112086002574.

Gomes, P. J., Sousa, R. G., Cardoso, J. I., Dias, M. M., & Lopes, J. C. B. (2007a). Studies in a large power transformer—Heat transfer and flow optimization using CFD. In *Advanced Research Workshop on Transformers,* Baiona (pp. 130–146).

Gomes, P. J., Sousa, R. G., Dias, M. M., & Lopes, J. C. B. (2007b). Large power transformer cooling—Flow simulation and PIV analysis in an experimental prototype. In *Advanced Research Workshop on Transformers,* Baiona (pp. 113–129).

IEC. (2005). IEC 60076: Power Transformers—Part 7: Loading guide for oil-immersed power transformers.

Kranenborg, E. J., Olsson, C. O., Samuelsson, B. R., Lundin, L. A., & Missing, R. M. (2008). Numerical study on mixed convection and thermal streaking in power transformer windings. In *5th European Thermal-Sciences Conference,* Netherlands.

Langhame, Y., Castonguay, J., Bedard, N., & St-Onge, H. (1985). Low temperature performance of naphthenic and paraffinic oils in transformers and automatic circuit reclosers. *IEEE Power Engineering Review PER-, 5,* 47. https://doi.org/10.1109/MPER.1985.5528827.

Montsinger, V. M. (1930). Loading transformers by temperature. *Transactions A.I.E.E, 776–790.*

Oliveira, R. T. (2014). *Modelling and simulation of flow and heat transfer in shell-type power transformers.* University of Porto.

Oliver, A. J. (1980). Estimation of transformer winding temperatures and coolant flows using a general network method. In *IEE Proceedings C - Generation, Transmission and Distribution.* IET (pp. 395–405). https://doi.org/10.1049/ip-c:19800061.

Picher, P., Torriano, F., Chaaban, M., Gravel, S., Rajotte, C., & Girard, B. (2010). Optimization of transformer overload using advanced thermal modelling. In *Cigré,* Paris (p. A2_305_2010).

Radakovic, Z. R., & Sorgic, M. S. (2010). Basics of detailed thermal-hydraulic model for thermal design of oil power transformers. *IEEE Transactions on Power Delivery, 25,* 790–802. https://doi.org/10.1109/TPWRD.2009.2033076.

Roshenow, W., Hartnett, J., & Cho, Y. (1998). *Handbook of heat transfer* (3rd ed.). McGraw-Hill.

Schmidt, N., Tenbohlen, S., Chen, S., & Breuer, C. (2013). Numerical and experimental investigation of the temperature distribution inside oil-cooled transformer windings. In *18th International Symposium on High Voltage Engineering,* Seoul, South Korea.

Shah, R. K., & London, A. L. (1971). *Laminar flow forced convection heat transfer and flow friction in straight and curved ducts. A summary of analytical solutions.* California: Stanford University, Mechanical Enginnering.

Skillen, A., Revell, A., Iacovides, H., & Wu, W. (2012). Numerical prediction of local hot-spot phenomena in transformer windings. *Applied Thermal Engineering, 36,* 96–105. https://doi.org/10.1016/j.applthermaleng.2011.11.054.

Tanguy, A., Patelli, J. P., Devaux, F., Taisne, J. P., & Ngnegueu, T. (2004). Thermal performance of power transformers: Thermal calculation tools focused on new operating requirements. In *Cigré,* Paris (p. A2_105_2004).

Torriano, F., Chaaban, M., & Picher, P. (2010). Numerical study of parameters affecting the temperature distribution in a disc-type transformer winding. *Applied Thermal Engineering, 30,* 2034–2044. https://doi.org/10.1016/j.applthermaleng.2010.05.004.

Weinlader, A., & Tenbohlen, S. (2009). Thermal-hydraulic investigation of transformer windings by CFD-modelling and measurements. In *Proceedings of the 16th International Symposium on High Voltage Engineering,* Johannesburg, South Africa.

Wu., W., Wang, Z. D., Revell, A., Iacovides, H., & Jarman, P. (2012). Computational fluid dynamics calibration for network modelling of transformer cooling oil flows—Part I: heat transfer in oil ducts. *IET Electric Power Applications, 6,* 19–27. doi:https://doi.org/10.1049/iet-epa.2011.0004.

Wu., W., Wang, Z. D., Revell, A., & Jarman, P. (2012). Computational fluid dynamics calibration for network modelling of transformer cooling flows—Part II: pressure loss at junction nodes. *IET Electric Power Applications, 6,* 18–24. doi:https://doi.org/10.1049/iet-epa.2011.0005.

Yatsevsky, V. A. (2014). Hydrodynamics and heat transfer in cooling channels of oil-filled power transformers with multicoil windings. *Applied Thermal Engineering, 63,* 347–353. https://doi.org/10.1016/j.applthermaleng.2013.10.055.

Chapter 5
FluSHELL Validation

Abstract In this chapter the FluSHELL tool is directly compared against both experiments and CFD. The CFD results have been compared against experiments in Chap. 3. The current comparison aims to validate FluSHELL temperature predictions and its underlying physical assumptions.

In this chapter the FluSHELL tool is directly compared against both experiments and CFD. The CFD results have been compared against experiments in Chap. 3. The current comparison aims to validate FluSHELL temperature predictions and its underlying physical assumptions.

In Sect. 5.1, the novel tool developed is compared with experimental data previously reported and compared with CFD in Chap. 3.

However, the exclusive comparison between the FluSHELL tool is not conclusive neither fair. In FluSHELL the coil/washer system has been considered to be fully adiabatic. Thus, this comparison has been complemented including a newly developed CFD model.

The geometry of this new CFD model is the same as used for the scale model validation in Chap. 3. Apart from the fact that in this new model the acrylic plates, the oil pools, the steel structures and the pressboard blocks do not exist. This new model, based on previous, has been generated for the exclusive purpose of validating more consistently the physical concepts implemented in the FluSHELL tool so these components needed to be removed.

This new 3D CFD geometry, its boundary conditions and some results are described in the Sect. 5.2. In Sect. 5.3—FluSHELL performance is complementary compared against this new CFD results

In Sect. 5.4 final considerations are included about the accuracy of the FluSHELL tool developed in this work.

© Springer International Publishing AG 2018

H. Campelo, *FluSHELL – A Tool for Thermal Modelling and Simulation of Windings for Large Shell-Type Power Transformers*, Springer Theses,

https://doi.org/10.1007/978-3-319-72703-5_5

5.1 FluSHELL Versus Experiments

In this section the FluSHELL predictions are compared against experiments and against the CFD results that have been validated in Chap. 3.

Figure 5.1 shows the comparison between the average temperatures of the turns predicted using FluSHELL and the measurements. The comparison is shown for the three sets of experiments conducted.

FluSHELL over predicts the average temperatures of the turns on average in circa 6 °C. It is also observed that FluSHELL predicts adequately the temperature trend both within the same set of experiments and also between each set of experiments. FluSHELL consistently predicts the highest temperatures for the set of experiments EXP1-EXP3, lower temperatures for the set EXP7-EXP9 and the lowest for the set EXP4-EXP6. This is the exact trend of the measurements. Although this direct comparison is not fair, because the physical assumptions implemented in the FluSHELL tool are not exactly reproduced in the experimental setup.

As an example, Fig. 5.2 shows the temperature maps in the coil for EXP1 conditions. Figure 5.2a depicts the solid temperature distribution predicted using FluSHELL and Fig. 5.2a the solid temperature distribution predicted using the CFD model presented in Chap. 3. It is noteworthy that these latter CFD results have been validated against the experiments, thus confidently represent the experimental setup.

This comparison indicates that FLUSHELL predicts significantly higher temperatures (>20 °C) at the innermost and outermost turns than those being predicted with CFD. This is even more evident in Fig. 5.2c after rescaling the CFD

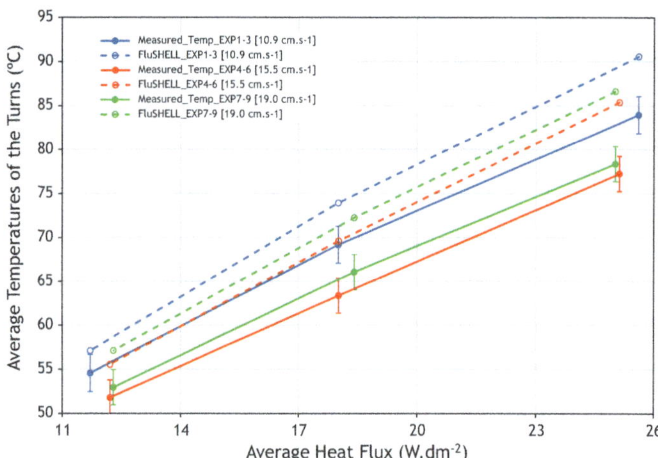

Fig. 5.1 Comparison between the average temperatures of the turns predicted with FluSHELL and measured (for all experiments)

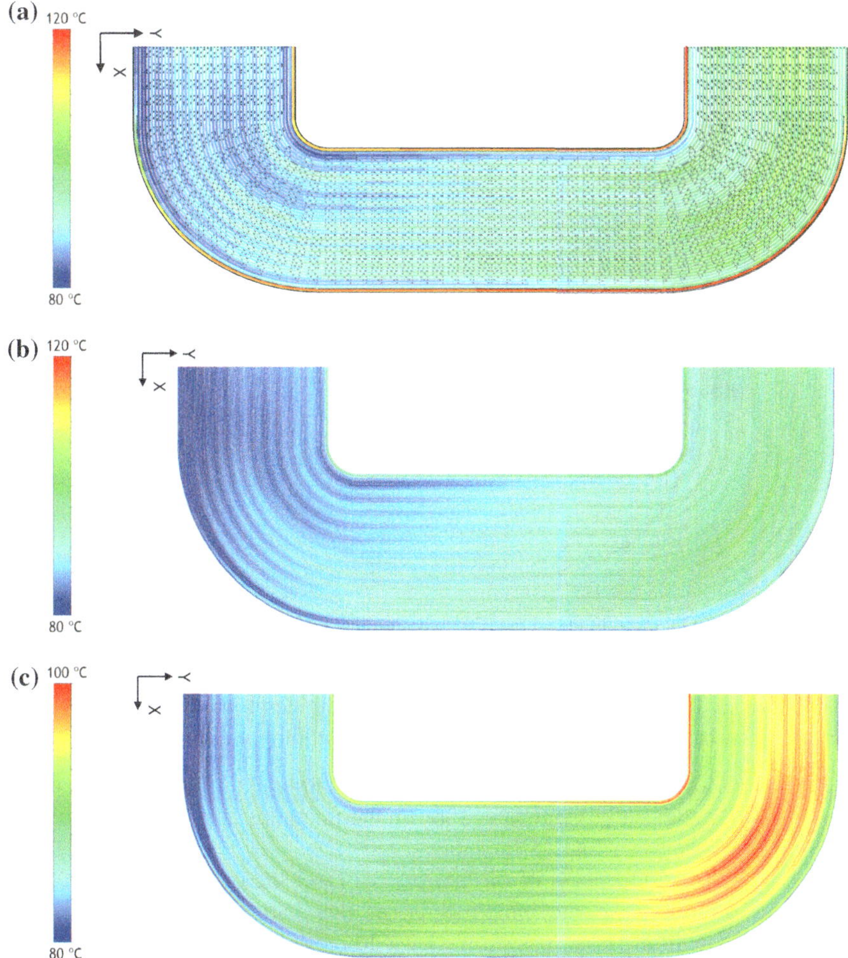

Fig. 5.2 Temperature maps in the coil for EXP1 conditions: **a** FluSHELL, **b** CFD Scale model and (c) CFD Scale model with a different temperature scale

temperature map between 80–100 °C. It is also noteworthy that for the remaining turns the temperature trend between FluSHELL and CFD seems similar, but CFD temperatures are consistently lower over the whole domain.

The same can be observed in the oil temperatures maps in Figs. 5.3a and b. Both plots show a hotter region of oil in the top curved region close to the outlet nozzle, however CFD exhibits colder oil temperatures more evident near the inlet and the outlet regions.

As mentioned above, the direct comparison between FluSHELL and experiments is neither fair nor possible. In fact the temperature differences discussed in

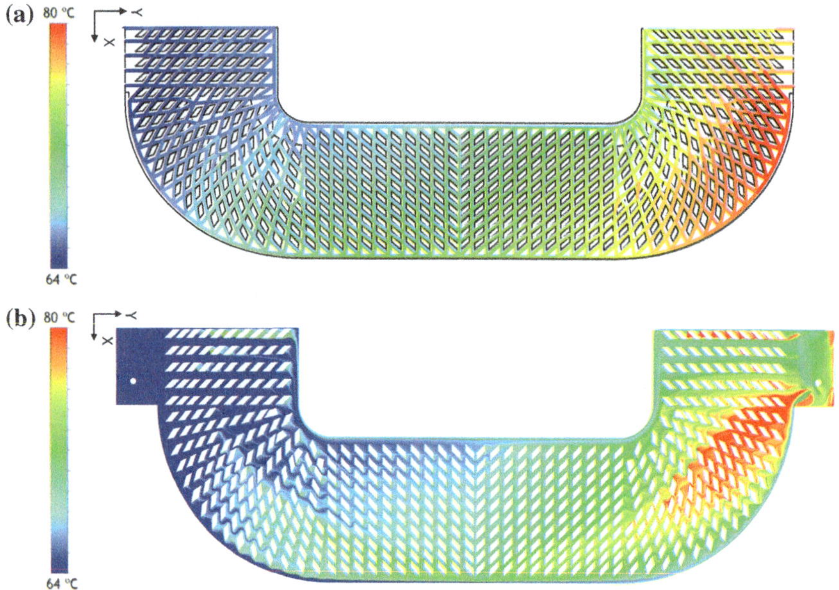

Fig. 5.3 Temperature maps in the oil for EXP1 Conditions: **a** FluSHELL and **b** CFD Scale Model

this section are related with a set of specific characteristics of the experimental setup that have not been implemented in the FluSHELL tool being them:

1. The pressboard blocks existing in the setup around the copper coil which decreases significantly the temperatures in the innermost and outermost turns—comparison between Figs. 5.2a and b;
2. The bottom and top pools of oil that partially wet the outermost turn in these regions and decrease the temperatures of this particular turn even further comparing with the innermost turn where this does not occur—Fig. 5.2c;
3. The acrylic plates existing in the setup which decreases the average temperature of the whole copper coil due to heat that is transferred to ambient air through these surfaces.

Although this set of specific characteristics are well known and have been clearly identified, there is no current evidence about its importance in commercial transformers. As a result, the most conservative approach is to consider the coil/washer system adiabatic.

For these reasons, an adequate 3D CFD model has been developed to be compared with FluSHELL and thus enable a rigorous conceptual validation of the tool. This validation is described in the next sections.

5.2 Adiabatic CFD Model

This adiabatic CFD model differs from both CFD models described in Chap. 3 and in Chap. 4.

The CFD model described in Chap. 3 includes all the components of the experimental setup. In that model the coil/washer system has been modelled together with all the surrounding components included in the setup such as the acrylic plates, the steel structures, the polystyrene plates, the pressboard blocks, the bottom and top oil pools. This model aimed at representing as closest as possible the conditions of the setup.

In addition, the CFD model described in Chap. 4 used to extract friction and heat transfer coefficients has been intentionally simplified for the purpose. In that model, the surrounding components do not exist either, but the turns have been approximated by a constant heat flux wall. Moreover, the spacers and the insulation frames have been neither been meshed nor fully represented.

Thus, an adiabatic 3D CFD model of the coil/washer has been created using exactly the same coil/washer dimensions of the model described in Chap. 3 and avoiding the extreme approximations of the model described in this chapter.

5.2.1 Geometry

The geometry shown in Fig. 5.4 and used to validate FluSHELL is based on the experimental setup described in Chap. 2.

Figure 5.4 illustrates the computational domain comprising two regions. These two regions are exactly the same as used in Chap. 3. Figure 5.4a shows a *XY* plane of the 3D region that includes the fluid volumes (white coloured) along with the spacers as well as the inner and outer insulation frames (all black coloured). Figure 5.4b shows a *XY* plane of the 3D region that includes the turns (orange coloured) and the pressboard between turns. According to both images a *XZ* symmetry plane has been considered.

This geometry has been created in ANSYS Design Modeler®. These two regions have been superimposed along the *Z* coordinate, as shown in Fig. 5.5, using grid interfaces.

The reference dimensions of the region depicted in Fig. 5.4a are shown in Fig. 5.6.

Figure 5.6a depicts the generic dimensions of the domain while Fig. 5.6b shows the main inner dimensions of the same domain. Figure 5.6b also shows that the inner and outer insulation frames used are linear shaped with 7 mm length, which means these frames do not allow the innermost and outermost turns to be in direct contact with the cooling fluid.

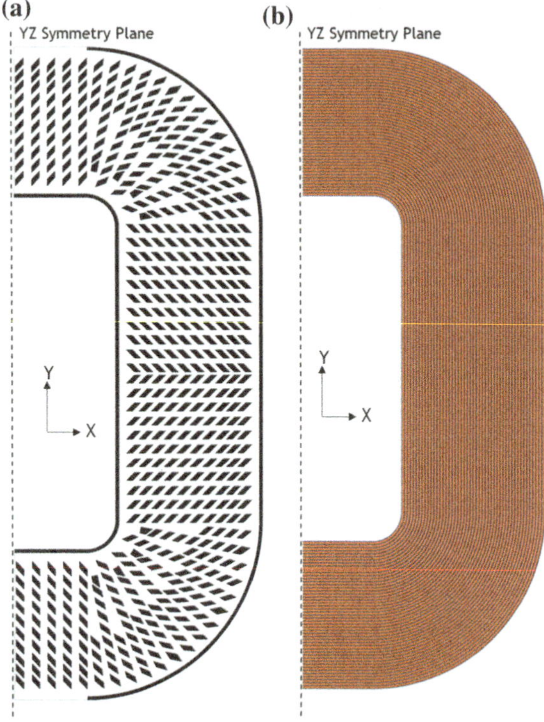

Fig. 5.4 Geometry of the adiabatic CFD model used for validating FluSHELL—**a** fluid region and **b** copper coil region

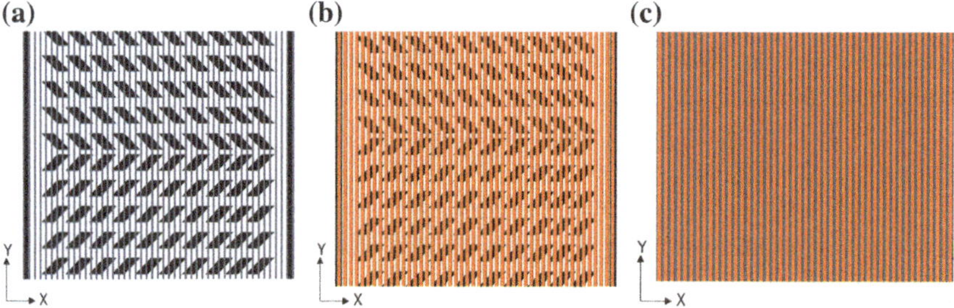

Fig. 5.5 Sequential superimposition of the regions—**a** pressboard between turns; **b** turns and **c** the final solid arrangement as considered

In a commercial shell-type transformer similar frames may be found in the innermost and outermost regions that may have different shapes (refer to Chap. 1). The linear shaped insulation frames herein considered have been designed to guarantee absence of heat convection in the surfaces of the innermost and outermost

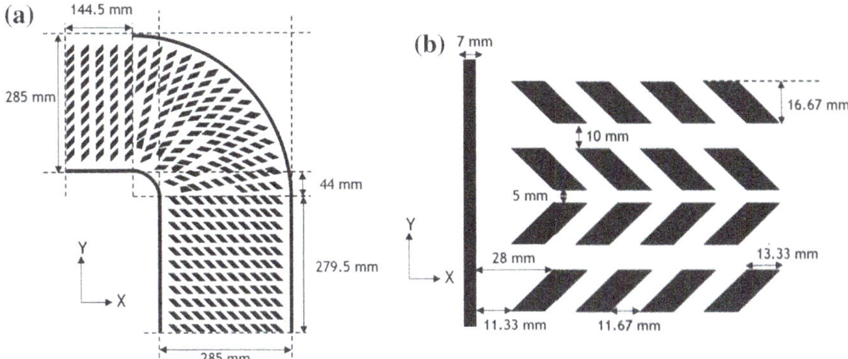

Fig. 5.6 Reference dimensions of the region of the domain identified in Fig. 5.4a—**a** external dimensions; **b** solid structures arrangement and dimensions

turns and hence enabling a deeper understanding of the heat transfer mechanisms in these locations where the hottest temperatures are expected.

The reference dimensions of the region depicted in Fig. 5.4b concerning the turns are shown in Fig. 5.7.

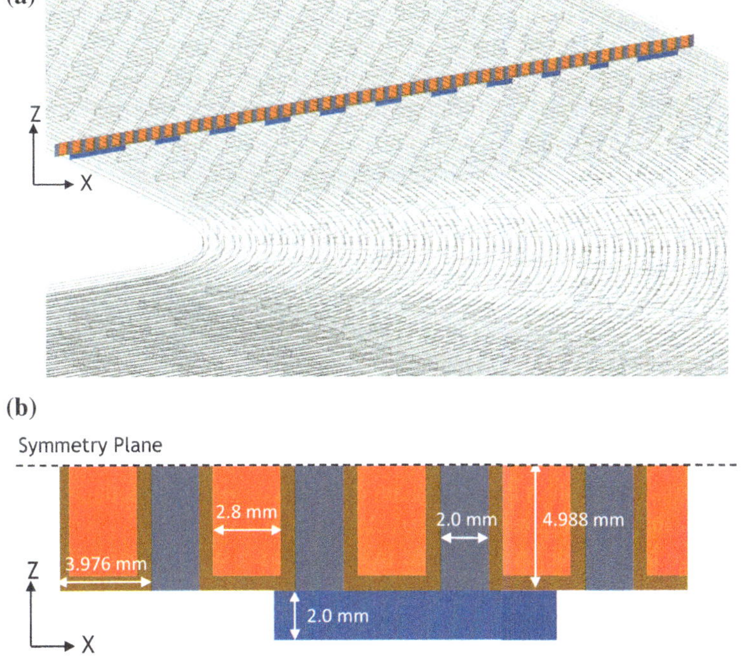

Fig. 5.7 Reference dimensions of the region of the domain identified in Fig. 5.4b—**a** cut view using XZ plane; (b) detailed arrangement and dimensions of the turns with an adjacent fluid channel

The height of the fluid region along the Z Coordinate is 2 mm (blue coloured in Fig. 5.7). As shown in Fig. 5.7b the turns are fully modelled without simplifications. The bulk of each turn is homogeneous comprising a single rectangular copper bar (with 2.8×8.8 mm) wherein the heat source is imposed. According to Fig. 5.7b only half height (4.4 mm) has been modelled as symmetry is assumed along the Z Coordinate. There are 48 turns equally distributed along the X Coordinate. Each copper bar is covered with 0.988 mm thickness of insulation paper. A uniform thickness of 2 mm of pressboard between each turn has been considered. An exception to this pattern occurs before the innermost turn and after the outermost turn, where a pressboard thickness of 1 mm has been considered. In commercial transformers this additional pressboard at the extremities is referred as *edge strips*.

5.2.2 Mesh

In the current model an unstructured mesh of 25.56 Million elements has been used. Contrarily to the CFD model in Chap. 4 used for calibration, the spacers, the insulation frames and the turns have also been meshed. Figure 5.8a shows the type and resolution of the mesh elements used in the vicinity of the inner insulation frame and Fig. 5.8b illustrates shows the resolution around the spacers.

The mesh depicted in Fig. 5.8a indicates an inflated layer of mesh elements near the hot surfaces of both the spacers and the insulation frames.

The mesh depicted in Fig. 5.8b also indicates that enough elements have been considered between each two consecutive walls, hence enabling an adequate modelling of the fluid velocity and temperature profiles even in the smallest fluid channels, which corresponds to channel illustrated with 14 cells across it. This small fluid channel is in the inner bottom curved region of domain (refer to Fig. 5.4a).

Complementary Fig. 5.9a shows the mesh resolution in the inner insulation frame along the Z Coordinate (a similar mesh has been applied to the outer insulation frame) while in Fig. 5.9b a XZ plane is used to detail the mesh used in the turns and in the adjacent fluid channels.

Figure 5.9a depicts the inner insulation frame meshed with 9 cells along X- coordinate and 9 cells along Z-coordinate. Figure 5.9b shows the mesh used in the turns and in an adjacent fluid channel. Due to the complexity of the domain, it is observable that the mesh in the fluid channel is not conformal with the mesh used in the turns. In the fluid region 9 layers of elements are inflated with a growth rate of 20% near the hot surface. As observed in the mesh sensitivity analysis included in Chap. 4 this mesh resolution in the fluid represents an adequate compromise between computational effort and accuracy.

The distribution of mesh elements among the different components of the domain is summarized in Table 5.1.

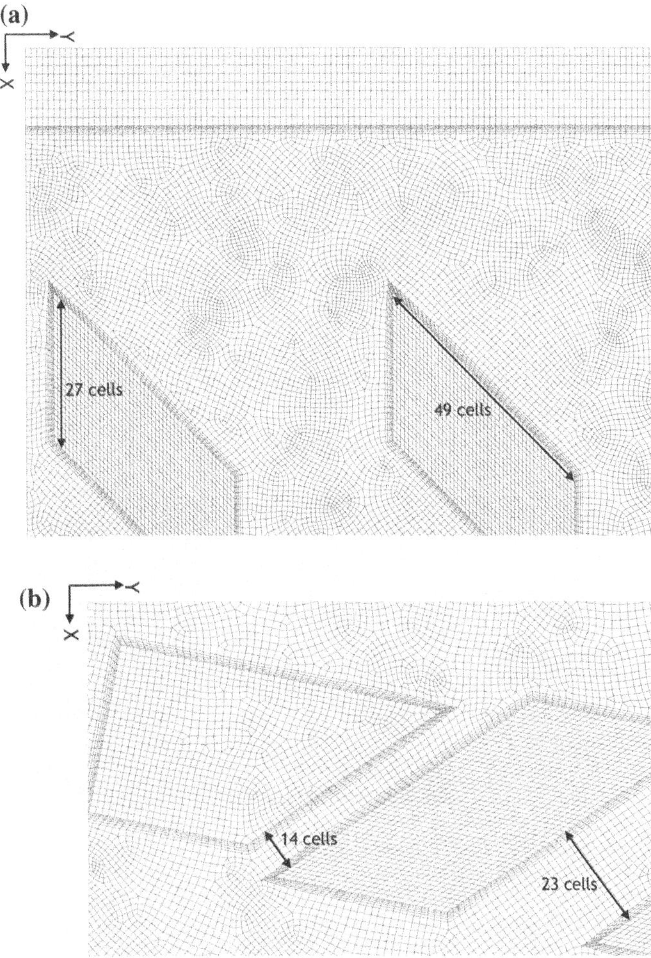

Fig. 5.8 Type of mesh elements and mesh resolution used—**a** in the spacers and **b** in the fluid regions surrounding the spacers

5.2.3   Boundary Conditions

The boundary conditions used in these CFD simulations are listed in Table 5.2. The simulations have been conducted using ANSYS Fluent® 16.2.

Continuity, energy and momentum equations for the three Cartesian components of the velocity have been computed. Numerical residuals of 10^{-6} for continuity and momentum equations and residuals of 10^{-9} for energy equation have been considered as a condition of convergence but not self-sufficient. Local velocities and temperatures in randomly chosen locations have been also monitored and observed to be varying, between iterations, in the same order of magnitude as the residuals.

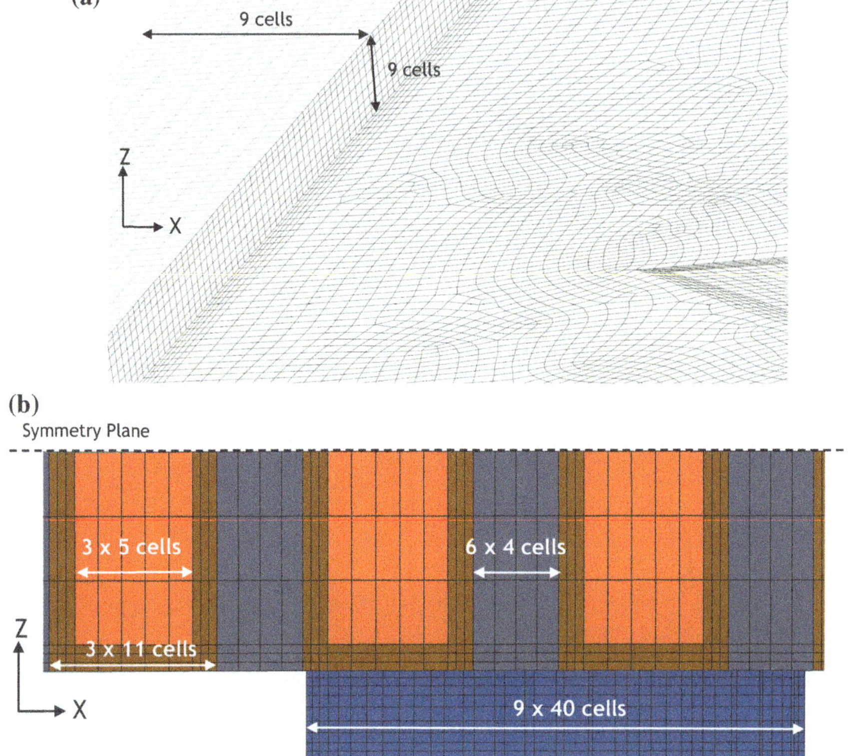

Fig. 5.9 Type of mesh elements and mesh resolution used along Z-coordinate – (a) in the inner insulation frame and (b) in the turns

Table 5.1 Distribution of the mesh elements between the different components of the domain

Component	MElements	% (out of 25.56 MElements)
Fluid	13.1	54.5
Spacers	6.1	24.0
Turns	5.1	20.0
Insulation frames	0.4	0.5

A total of 9 CFD simulations have been conducted. The 9 simulations correspond to the 9 experiments reported comprehensively in Chap. 3. The most relevant operating conditions of the experiments, used as boundary conditions in these CFD simulations, are listed in Table 5.3.

Table 5.3 identifies the range of total inlet pressures used in the simulations and the corresponding average fluid velocities $\overline{u_G}$. The same heat generation rate has been imposed in each turn. Similarly, the heat source imposed in the turns is shown

Table 5.2 Boundary conditions and most relevant solver parameters

Inlet	Total pressure	Variable and uniform. Tables 3.3 and 5.3
	Temperature	Variable. Tables 3.3 and 5.3
Outlet	Static pressure	0 Pa
Turns	Volumetric heat source	Variable W m^{-3}. Tables 3.3 and 5.3
	Momentum	No-slip condition
Spacers	Volumetric heat source	0 W m^{-3}
	Momentum	No-slip condition
Insulation frames	Volumetric heat source	0 W m^{-3}
	Momentum	No-slip condition
External surfaces	Momentum	No-slip condition
	Heat transfer	Adiabatic
Symmetries	XY plane at middle height of the turns (Fig. 5.7b)	
	XZ plane (Fig. 5.4)	
Solver	Pressure based solver	
	Pressure-velocity coupling (CFL = 10 and URF = 1.0)	
	2nd order discretization schemes	
	Gradients with least square cell based	

Table 5.3 Inlet conditions and volumetric heat sources used as boundary conditions in the adiabatic CFD simulations

Sim. ID	Total inlet pressure [Pa]	Fluid inlet Temp. [°C]	$\bar{u}_G$ [cm.s^{-1}]	Turns heat source [kW m^{-3}]	Average heat flux [W dm^{-2}]
EXP1	3445	63.2	10.5	842.05	25.6
EXP2	4045	64.1	10.3	592.37	18.0
EXP3	6140	44.0	12.1	399.06	11.7
EXP4	5909	61.8	15.4	826.43	25.1
EXP5	6905	51.9	14.8	593.03	18.0
EXP6	9551	43.6	16.3	400.32	12.2
EXP7	7443	64.8	18.7	823.57	25.0
EXP8	8905	55.7	19.0	606.19	18.4
EXP9	11305	45.7	19.2	404.66	12.3

along with the resulting average heat flux obtained dividing the total heat generated by the wetted surfaces of the turns and of the pressboard between turns. Both averaged quantities are relevant design parameters and the ranges listed in Table 5.3 correspond to common design practices.

The relevant physical properties of the cooling fluid considered in these simulations are summarized in Table 5.4.

Table 5.4 Physical properties of the cooling fluid as implemented in the adiabatic CFD simulations

Type of fluid (Commercial reference)	Properties
Naphtenic mineral oil (Nynas Taurus®)	Dynamic viscosity, μ [kg m^{-1} s^{-1}]
	$\mathrm{Exp}((-20.04413) + (12078.37) \times (1/T(K))$ $+ (-4122209) \times (1/T(K))^2$ $+ (574840600) \times (1/T(K))^3)$
	Density, ρ [kg m^{-3}]
	$1065.801 - 0.6585 \times T(K)$
	Specific heat capacity, Cp [J kg^{-1} °C^{-1}]
	2016
	Thermal conductivity, k [W m^{-1} °C^{-1}]
	0.126

The dynamic viscosity and density expressions have been implemented using interpreted User Defined Functions. The expressions have been truncated for temperature values below 273.15 K and above 473.15 K.

The mineral oil herein used has the same commercial reference used in the other CFD models. As in Chap. 3, the dynamic viscosity and density values correspond to the latest data available from the supplier concerning the particular stock used in the experiments. The exception in this case is the specific heat capacity and thermal conductivity of the fluid which have been considered constant as in FluSHELL the same is assumed.

The solid components and the thermal properties of the corresponding materials conductivities are listed separately in Table 5.5.

As the CFD models have been conducted in steady-state the remaining physical properties of the solids are not relevant.

5.2.4 Results

The main CFD results for EXP1 conditions are presented here. The flow and temperature patterns obtained with this adiabatic model are compared against the

Table 5.5 Materials and corresponding thermal conductivities of the materials considered in the solid components of the domain

	Material	Thermal conductivity [W m^{-1} °C^{-1}]
Copper coil	Copper	388.5
	Kraft paper	0.16
	Pressboard	0.16
Insulation frames	Pressboard	0.16
Spacers	Acrylic	0.035

equivalent patterns obtained in CFD model described in Chap. 3. Figure 5.10 shows the velocity magnitude map at the middle height of the fluid domains of both models.

The comparison between Figs. 5.10a and b shows that the bottom oil pool induces somewhat different velocity conditions at the inlet of the washer. In the latter case, the inlet nozzle located in the bottom oil pool seems to be forcing more oil to enter the washer near the curved region than the case where the domain is truncated and uniform pressure conditions are applied. Although, the flow patterns inside the washer are still quite similar with fluid flowing preferentially near the insulation frames.

Figure 5.11 shows the fluid temperature map over the same plane and for both CFD models and for the same EXP1 conditions.

The temperature map in Fig. 5.11a shows the same temperature pattern when comparing with the map in Fig. 5.11b. The difference is that the oil temperatures are higher since, in this adiabatic model, the oil and the spacers are not transferring heat to the ambient air through the acrylic plate.

Figure 5.12 shows the solid temperature map at the *XY* symmetry plane cutting the copper coil in two halves for the same EXP1 conditions and for both CFD models.

Figure 5.12 shows that truncating the domain results in a similar relative temperature trend with much higher temperatures in the innermost and outermost turns

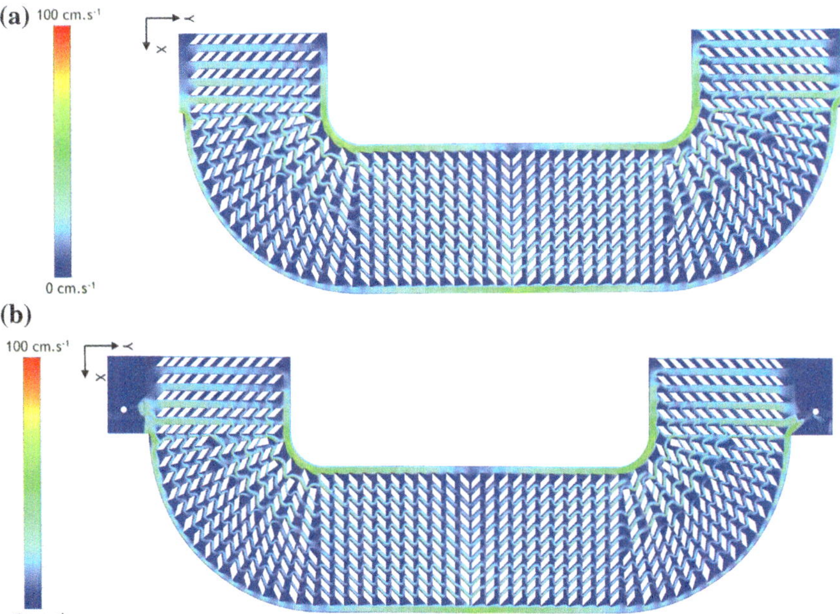

Fig. 5.10 Velocity magnitude map for EXP1 simulation in a plane located at middle height of the fluid channels (Z = 0.001 m): **a** adiabatic CFD model and **b** CFD model from Chap. 3

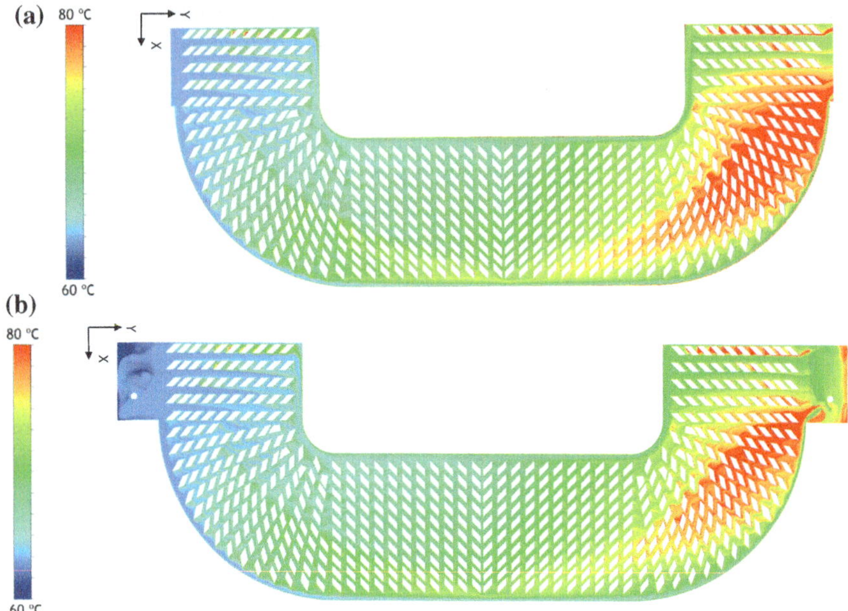

Fig. 5.11 Temperature map for EXP1 simulation in a plane located at middle height of the fluid channels (Z = 0.001 m): **a** adiabatic CFD model and **b** CFD model from Chap. 3

(>20 °C). Two main characteristics of the CFD domain described in Chap. 3 and depicted again in Fig. 5.12b seems to explain this:

1. The existence of pressboard blocks around the copper coil. Although made of a low thermal conductivity material the existence of such components is a relevant degree of freedom for these turns and thus it lowers considerably the maximum temperature observed;
2. The bottom and top oil pools are extended up to 45° in the curved regions. In this extended curved region up to 45° the oil is in direct contact with the outermost coil thus creating an additional degree of freedom. This further explains why the outermost turn is observed to have lower temperatures than the innermost turn.

Figure 5.13 includes the solid temperature map in the same plane at the middle height of the fluid domain of the adiabatic CFD model.

The results in Fig. 5.13a and b show significant temperature gradients in the insulation frames and in the spacers. Gradients of ~ 10 °C in the spacers and ~ 30 °C in the insulation frame are observed.

The same gradients, from a different perspective are shown in Fig. 5.14 over two XZ cut view planes crossing simultaneously the copper coil and the adjacent fluid domain.

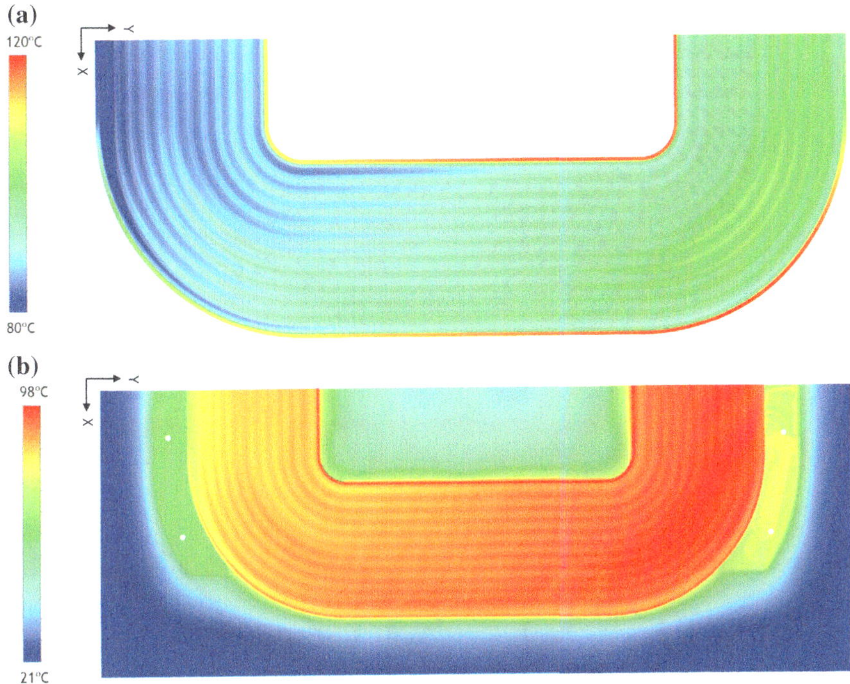

Fig. 5.12 Temperature maps for EXP1 simulation in the XY symmetry plane cutting the copper coil (Z = 0.006988 m): **a** adiabatic CFD model and **b** CFD model from Chap. 3

Figures 5.13and 5.14a, b show that, although made of low thermal conductivity materials, these structures are expected to contribute with non-negligible heat fluxes to the fluid.

In EXP1 simulation the total heat generated in the turns is 737.15 W and Table 5.6 identifies the surfaces through which this energy is transferred to the oil.

As expected, the main surfaces transferring heat to the fluid are the wetted surfaces of the turns. However, these surfaces transfer 69% of the total heat generated with a non-negligible amount of 29% being transferred through the pressboard and through the spacers. Hence, ignoring these structures in the modelling approaches may result in a significant overestimation of temperatures.

Furthermore, the remaining 2% of the total heat generated is being transferred to the oil through the wetted surfaces of the inner and outer insulation frames. Even though globally low, the amount of heat transferred through these surfaces is quite relevant for the innermost and outermost turns representing 47.9 and 38.1% of the heat being generated locally. As these turns are not in direct contact with oil and also due to the adiabatic domain, the remaining degree of freedom is to transfer heat to the neighbouring turns.

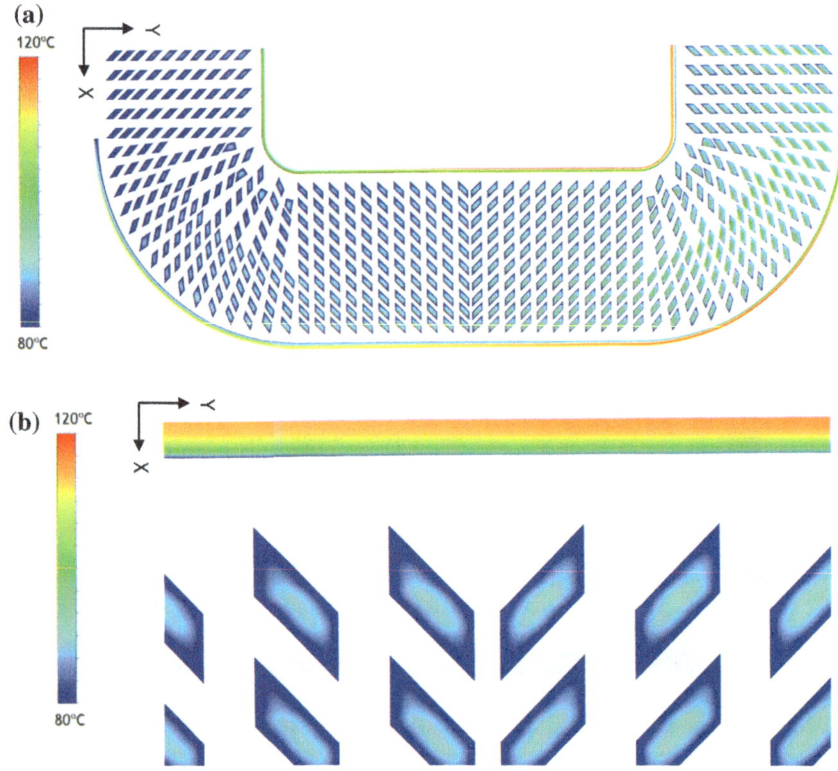

Fig. 5.13 Planes located at middle height of the fluid channels (Z = 0.001 m). Temperatures in the spacers and in the insulation frames: **a** normal view and **b** zoomed view

5.3 FluSHELL Versus Adiabatic CFD

FluSHELL and CFD simulations have been conducted for the 9 experimental conditions referred in Table 3.3. The same mineral oil has been considered in both numerical approaches according to the physical properties listed in Table 3.4. The same materials have also been considered for the solid components of the domain according to Table 3.5.

For validation purposes the same average fluid velocity across G section, $\overline{u_G}$, has been imposed as a boundary condition (Table 3.3). This means the resulting oil mass flow rates are practically the same in both FluSHELL and CFD, exhibiting maximum differences lower than 2.1% for each simulation. Table 5.7 summarizes the global characteristics of the 9 simulations conducted.

The white shaded columns in Table 5.7 list direct or indirect boundary conditions and the grey shaded columns list the global results useful to compare FluSHELL and CFD simulations. These global results are:

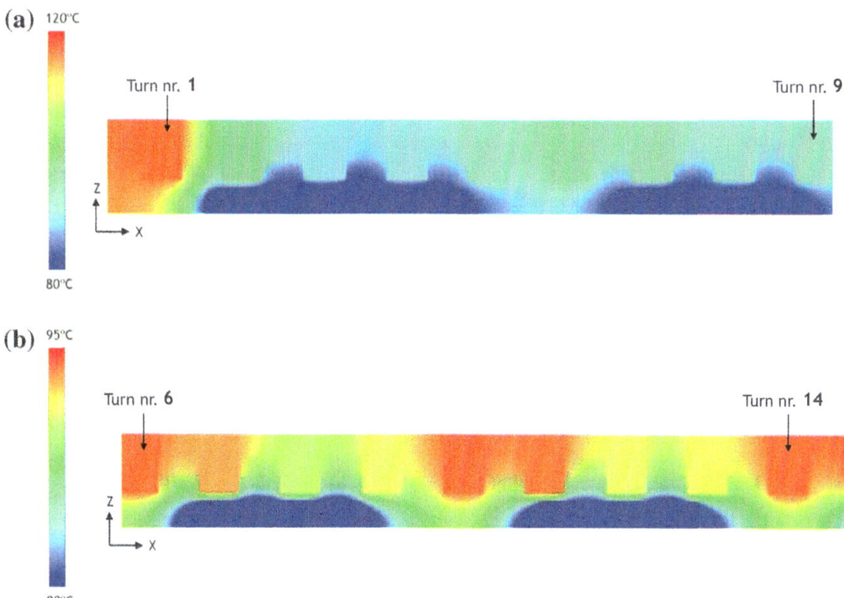

Fig. 5.14 Temperature maps for EXP1 simulation in a XZ plane located at Y = 0.66682 m. Temperatures in the copper coil, adjacent fluid channels and remaining solid structures: **a** from Turn nr. 1 to Turn nr. 9 and **b** from Turn nr. 6 to Turn nr. 14

Table 5.6 Heat transferred to the oil across each component of the domain (for EXP1 simulation)

		Heat [W]	Local heat [W]/Total [W]
Total heat generated		737.15	100%
Transferred from	To		
Inner insulation frame	Fluid	5.14	≈1% (47.9% of 10.7 W)
Outer insulation frame	Fluid	7.58	≈1% (38.1% of 19.9 W)
Turns	Fluid	507.97	69
Pressboard between turns	Fluid	126.22	17
Spacers	Fluid	90.24	12

- **the outlet oil temperature,** which represents the mass weighted average temperature at the outlet surfaces. As the domain is closed and its boundaries adiabatic, this temperature corresponds to the sensible enthalpy variation within the oil. Results shows that the average deviation between FluSHELL and CFD is 0.1 °C, which is considered acceptable.
- **the residuals**, on the mass and energy balances after the iterative calculation loops have been considered to converge. The mass flow rate at the outlet must

Table 5.7 Global characteristics of FluSHELL and CFD simulations used for validation purposes

SIM. ID	Numerical model	Inlet toil [C]	Outlet toil [C]	Mass flow [kg s^{-1}]	Source Heat [W]	Residuals Mass [kg s^{-1}]	Energy [W]	Sim. Time [min]
EXP1	CFD	63.2	77.9	2.49E-02	737.15	9.00E-07	9.00E-07	243
	FluSHELL	63.2	77.9	2.49E-02	737.15	8.00E-14	2.00E-04	2.8
EXP2	CFD	54.1	64.7	2.43E-02	518.58	9.00E-07	9.00E-07	212
	FluSHELL	54.1	64.5	2.47E-02	518.57	4.00E-14	2.00E-04	2.5
EXP3	CFD	44.0	50.3	2.76E-02	349.38	9.00E-07	9.00E-07	254
	FluSHELL	44.0	50.1	2.82E-02	349.35	6.00E-14	3.00E-04	2.1
EXP4	CFD	61.8	71.7	3.61E-02	723.49	9.00E-07	9.00E-07	266
	FluSHELL	61.8	71.6	3.66E-02	723.48	1.00E-13	2.00E-04	2.3
EXP5	CFD	51.9	59.2	3.52E-02	519.16	9.00E-07	9.00E-07	408
	FluSHELL	51.9	59.1	3.55E-02	519.15	7.00E-14	2.00E-04	2.3
EXP6	CFD	43.6	48.1	3.87E-02	350.44	9.00E-07	9.00E-07	204
	FluSHELL	43.6	48.0	3.94E-02	350.45	2.00E-13	3.00E-04	1.8
EXP7	CFD	64.8	73.0	4.37E-02	720.97	9.00E-07	9.00E-07	292
	FluSHELL	64.8	72.8	4.44E-02	720.98	1.00E-13	1.00E-04	2.4
EXP8	CFD	55.7	61.6	4.45E-02	530.68	9.00E-07	9.00E-07	258
	FluSHELL	55.7	61.5	4.55E-02	530.67	5.00E-14	2.00E-04	2.2
EXP9	CFD	45.7	49.6	4.55E-02	354.24	9.00E-07	9.00E-07	219
	FluSHELL	45.7	49.5	4.63E-02	354.25	6.00E-14	2.00E-04	1.7

equal the mass flow rate at the inlet and the enthalpy change within the fluid, in steady-state, must balance the energy injected into the system through the turns. Results in Table 5.7 show that both FluSHELL and CFD exhibit negligible differences between the expected and obtained values.

The dark grey shaded column represents a performance indicator based on **the time** needed to achieve converged solutions. FluSHELL is two orders of magnitude faster than CFD, with FluSHELL simulation time representing 0.9% of the average CFD simulation time.

These simulation times refer to an HP Workstation with 16 processors Intel Xeon$^®$ CPU E5-2680 @ 2.70 GHz and 256 GB of RAM. As described in Chap. 4, FluSHELL is currently coded in VBA for Microsoft Excel$^®$ and thus 1 single core has been used. The CFD simulations were run using 4 ANSYS HPCs$^®$ distributed over 8 cores.

The maximum and average temperatures predicted by FluSHELL and CFD for the 9 simulations are plotted in Fig. 5.15. The 9 simulations have been combined into three groups of simulations which share identical average fluid velocities.

For the range of heat fluxes simulated, FluSHELL tends to under predict the average temperature in circa 1.8 °C and to over predict the maximum temperature in 2.4 °C across the range of heat fluxes. The global trend is in good agreement with CFD and the observed temperature differences are close to the total combined

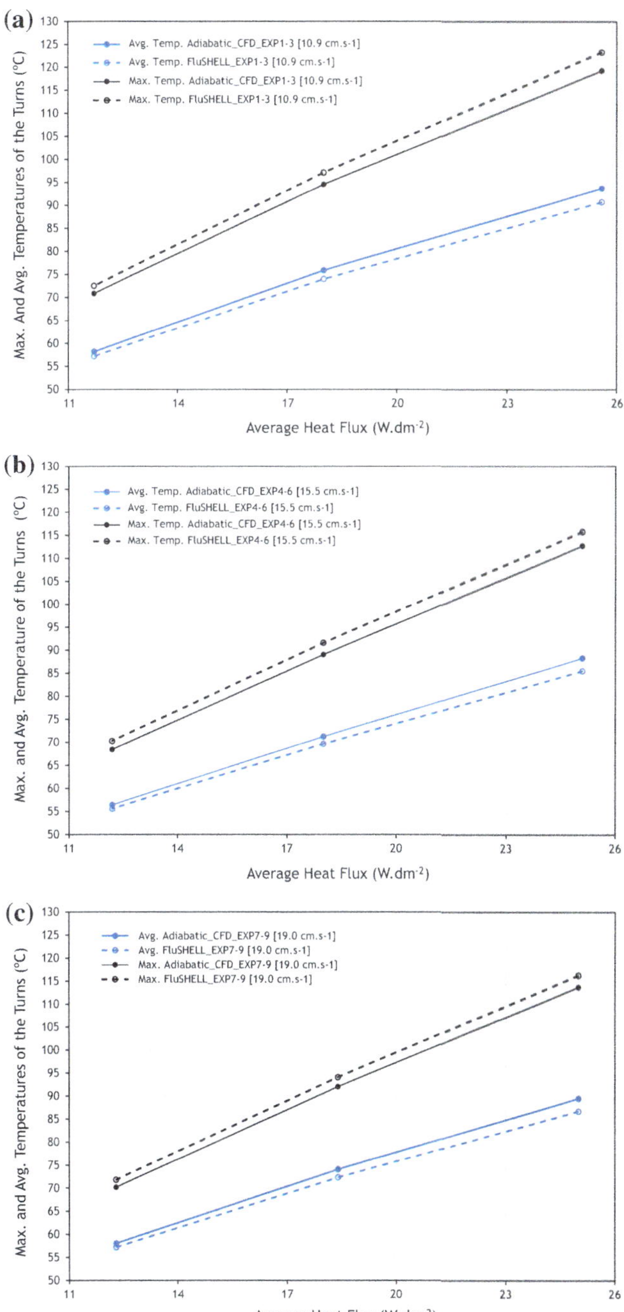

Fig. 5.15 Maximum and average temperatures of the turns predicted using FluSHELL and CFD —**a** EXP1-3; **b** EXP4-EXP6 and **c** EXP7-EXP9 simulations

uncertainties associated to the current measuring capabilities (refer to the combined uncertainties of the average copper coil temperature described in Chap. 2 and reported in Chap. 3).

The analysis proceeds with a more detailed comparison of the temperatures predicted in each turn. For this purpose, each turn has been numbered sequentially according to Fig. 5.16.

Results for the average and maximum temperatures observed in each turn are plotted in Fig. 5.17 through Fig. 5.19 and afterwards summarized in Table 5.8.

Figure 5.17 shows that the highest average and maximum temperatures occurs consistently in the turns in direct contact with both inner and outer insulation structures: Turn 1 and Turn 48. In the remaining turns, subtle temperature oscillations are observed due to the presence of the spacers and due to the different flow conditions across each turn. FluSHELL is observed to model adequately the hottest temperatures occurring in the innermost and outermost turns as well as the oscillatory behaviour in between these extremities. In fact, FluSHELL and CFD curves are parallel for all simulations herein conducted.

Compared with CFD simulations, for EXP1 to EXP3, FluSHELL predicts average temperatures with an average absolute deviation of 2.1 °C and maximum temperatures with an average absolute deviation of 2.7 °C.

Similar behaviour is observed for cases EXP4 to EXP9 as shown in Figs. 5.18 and 5.19. FluSHELL predicts average temperatures with an average absolute deviation of 1.9 °C and maximum temperatures with an average absolute deviation

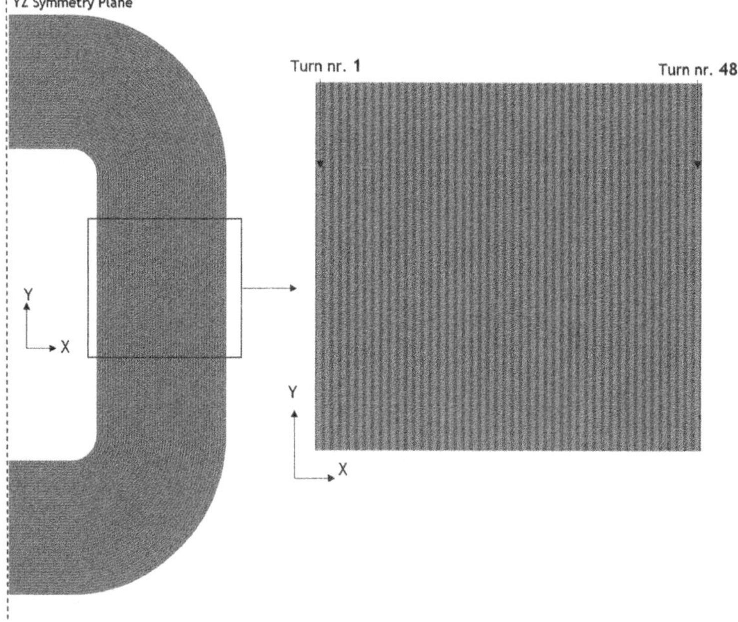

Fig. 5.16 Numbered turns

Fig. 5.17 a Average and
b Maximum predicted
temperatures for each turn.
EXP1-EXP3 simulations

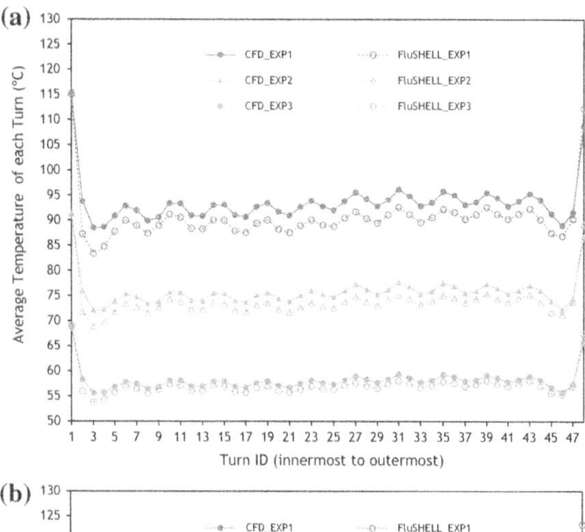

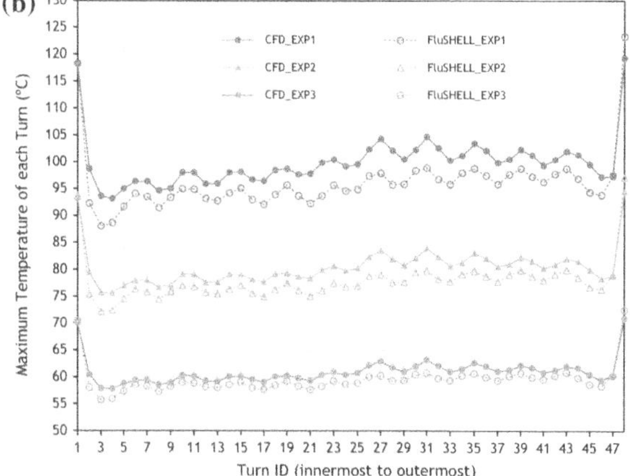

of 2.5 °C. These differences seem to increase for higher heat fluxes but are weakly
dependent on the mass flow rate.

Figure 5.20 shows the temperature distribution in the coil for EXP1 conditions
predicted using FluSHELL and CFD. The temperature maps in CFD are obtained
through interpolations over significant higher numbers of mesh elements, which
produces expectably smoother contours. However, both temperature maps are
identical.

To compare the local temperatures in different fluid channels as well as the mass
flow rate distribution, several control surfaces were created both in FluSHELL and
CFD. These surfaces are symmetrically distributed over the whole fluid domain
according to Fig. 5.21.

Figure 5.21a shows the control surfaces distributed along 2 dashed lines in the
top region of the domain which are identified as *Achannels* and numbered
sequentially according to the direction indicated by the arrows. Figure 5.21b shows
more control surfaces distributed along 3 dashed lines in the middle region of the

Table 5.8 Summary of the CFD and FluSHELL temperature predictions for EXP1-EXP9 simulations

SIM. ID	Turns CFD		Turns FluSHELL	
	Average Temp. [°C]	Max. Temp. [°C]	Average Temp. [°C]	Max. Temp. [°C]
EXP1	93.6	119.3	90.6	123.3
EXP2	75.9	94.5	74.0	97.1
EXP3	58.1	70.9	57.1	72.5
EXP4	88.2	112.8	85.4	115.8
EXP5	71.3	89.1	69.6	91.6
EXP6	56.4	68.4	55.6	70.3
EXP7	89.4	113.7	86.7	116.3
EXP8	74.0	92.0	72.3	94.1
EXP9	58.0	70.2	57.2	71.8

Fig. 5.18 a Average and **b** Maximum predicted temperatures for each turn. EXP4-EXP6 simulations

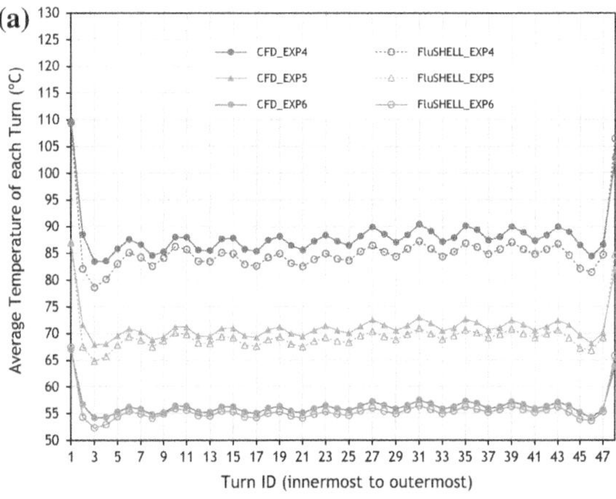

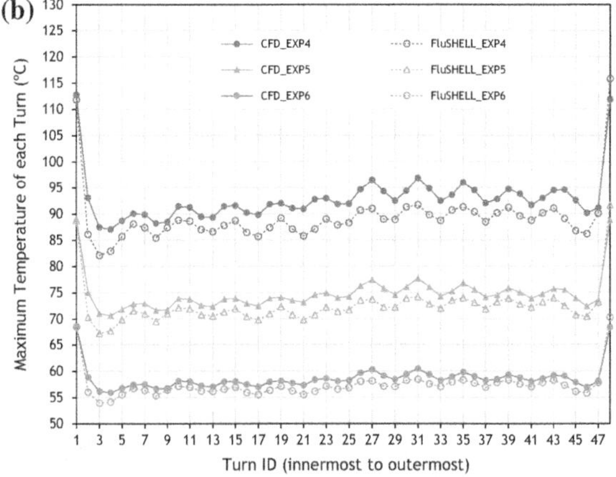

Fig. 5.19 **a** Average and **b** Maximum predicted temperatures for each turn. EXP7-EXP9 simulations

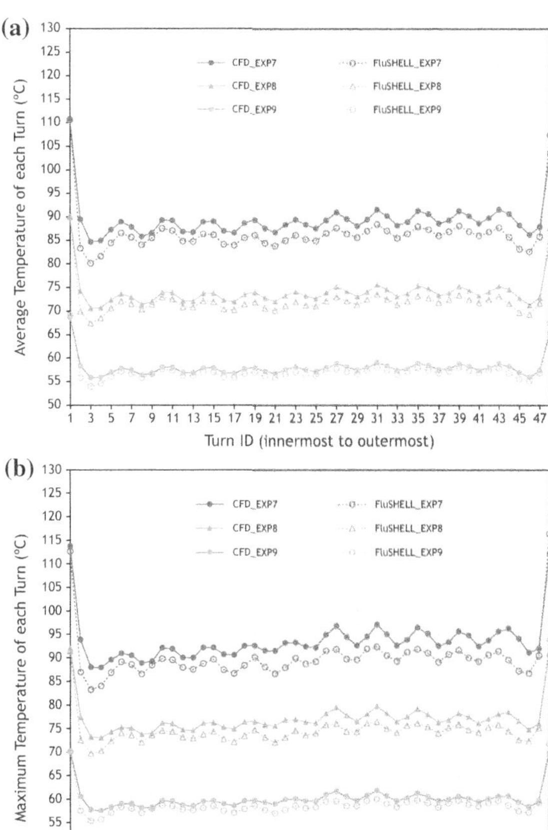

domain which are identified as *Gchannels* and numbered according to the arrows. Figure 5.21c shows the remaining control surfaces distributed along 2 dashed lines in the bottom region of the domain which are identified as *Bchannels* and numbered according to the arrows. There are 22 *Bchannels*, 33 *Gchannels* and 22 *Achannels* which amount for a total of 77 fluid channels. The total number of fluid channels in this geometry is 1268, representing approximately 6% of that total number and being distributed over representative regions of the domain.

Figure 5.22 shows a plot of the fluid temperatures observed, in both codes for EXP1 simulation. In this plot a series labelled *ALLchannels* has been used to group in the same series the *Bchannels*, *Gchannels* and *Achannels* numbered sequentially from bottom to top and ordered according to the directions identified in Fig. 5.21.

Figure 5.22 shows better agreement between FluSHELL and CFD in the lower regions of the domain (*Bchannels*) rather than the upper (*Achannels*), but even in the *Achannels*, corresponding to IDs from 56 up to 77, the average FluSHELL

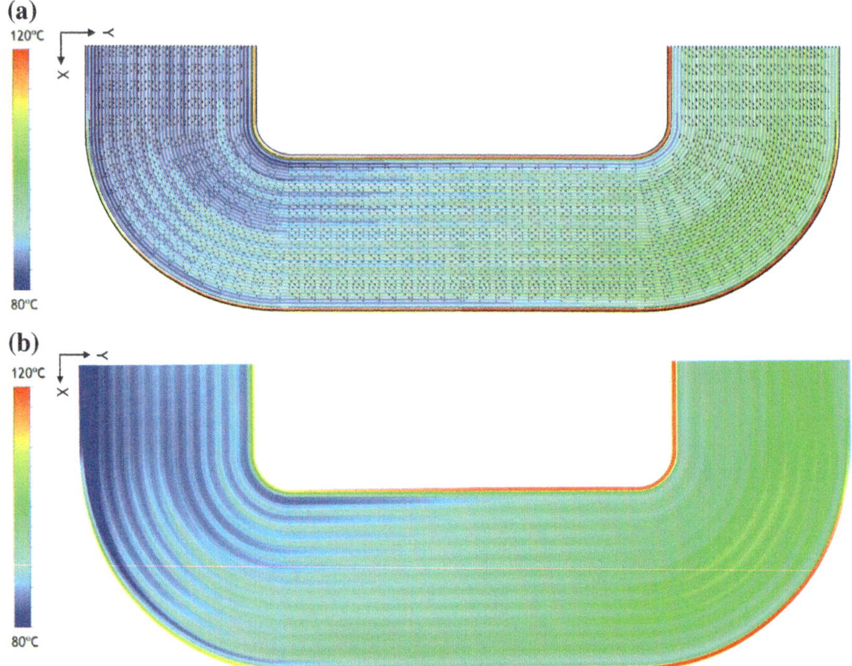

Fig. 5.20 Temperature maps in the coil for EXP1 conditions: **a** FluSHELL and **b** CFD

deviation to CFD is 1.3 °C. Figure 5.22 also shows that FluSHELL and CFD exhibit a similar fluid temperature behaviour as the fluid flows towards the outlet.

Comparison of FluSHELL and CFD predictions for simulations EXP1 to EXP9 are summarized in Table 5.9 where the deviations observed are consistently lower than 1.3 °C.

Figure 5.23 depicts, for EXP1, the mass flow rate fraction X_F in each of the 77 fluid channels, showing that the oil flows preferentially in the innermost and outermost channels next to the inner and outer insulation frames, respectively. An identical flow pattern is observed in the CFD velocity magnitude maps shown in Fig. 5.10 and this is a consequence of the higher flow area available in these specific channels. An exception to this pattern occurs in fluid channel 12 in Fig. 5.23a and fluid channel 1 in Fig. 5.23c, which correspond to the bottom and top curved regions of the domain.

Good agreement is observed over the whole domain. The deviations between the mass flow rates predicted using FluSHELL and CFD are summarized in Table 5.10.

Table 5.10 shows that the deviations are weakly dependent on the simulation conducted, which means the deviations are weakly dependent of the total mass flow rate imposed. In average, FluSHELL can predict the mass flow rate distribution with a deviation of 20% from CFD.

Fig. 5.21 Control surfaces created to compare mass flow rates and fluid temperatures— **a** Achannels; **b** Gchannels and **c** Bchannels

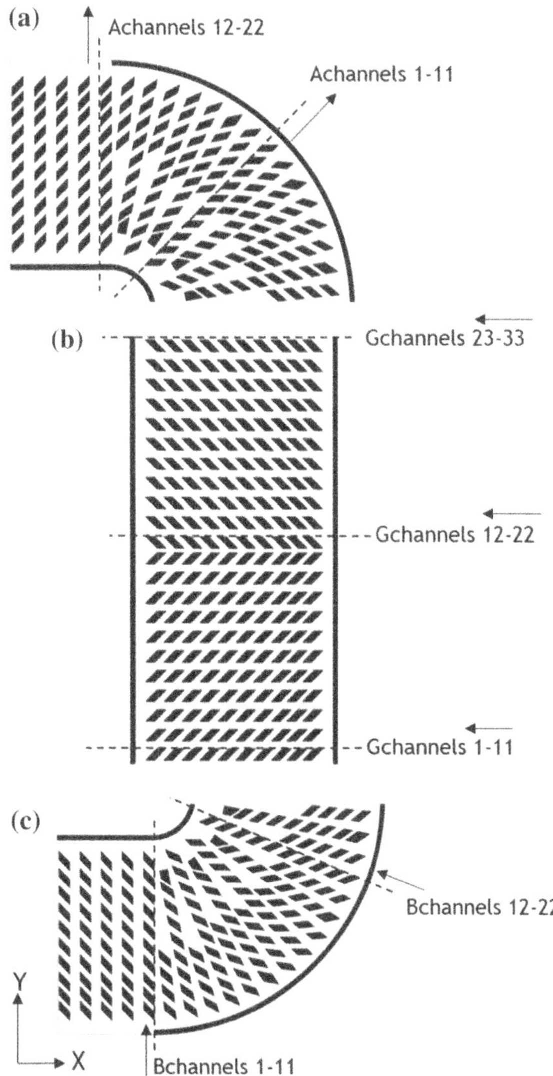

Table 5.11 list the deviations between the predicted pressure drops using FluSHELL and CFD.

FluSHELL over predicts the pressure in the coil/washer system on average in 20%. The deviations seem to be lower for higher oil velocities and for higher oil temperatures. The pressure predictions are consistently higher than CFD and the deviation is coherent with the relative deviation on the mass flow rates.

This is mainly due to differences in the characteristic length, L_{ch}, considered in FluSHELL to compute the frictional losses. As described in Chap. 4 the

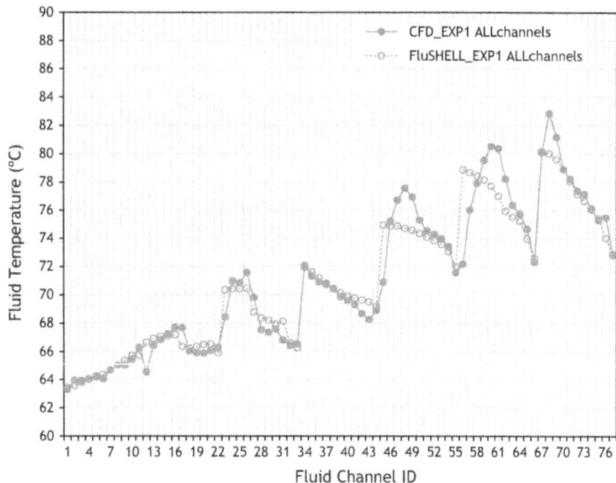

Fig. 5.22 Fluid temperature in the control fluid channels for EXP1. CFD and FluSHELL

Table 5.9 Fluid temperature deviations between FluSHELL and CFD

	Fluid temperature deviations— $\left| T_{ch}^{fn,FluSHSELL} - T_{ch}^{mwa,CFD} \right|$								
	EXP1	EXP2	EXP3	EXP4	EXP5	EXP6	EXP7	EXP8	EXP9
Bchannels	0.4	0.3	0.2	0.3	0.2	0.1	0.3	0.2	0.1
Gchannels	0.8	0.6	0.3	0.7	0.5	0.3	0.6	0.4	0.3
Achannels	1.3	0.9	0.5	1.3	0.7	0.4	1.3	0.8	0.4
Average	**0.9**	**0.6**	**0.3**	**0.7**	**0.5**	**0.3**	**0.7**	**0.5**	**0.3**

characteristic length of each fluid channel is calculated according to the schematic representation in Fig. 5.24.

FluSHELL considers the distance between vertex 1 and 4 as an adequate characteristic of the flow (Fig. 5.24a). However, between vertexes 1 and 4 the walls of the polygon are not all wetted (Fig. 5.24b). So, in practice, the pressure results in Table 5.11 seem to be indicating that this characteristic length results in a cumulative overestimation of the frictional losses.

Although, these pressure deviations are considered normal for these numerical techniques, and this still need to be assessed over a wider range of coil/washer geometries before refining the current approach. This is certainly a topic of future interest.

Fig. 5.23 Relative mass flow rate distribution for EXP1 using both FluSHELL and CFD

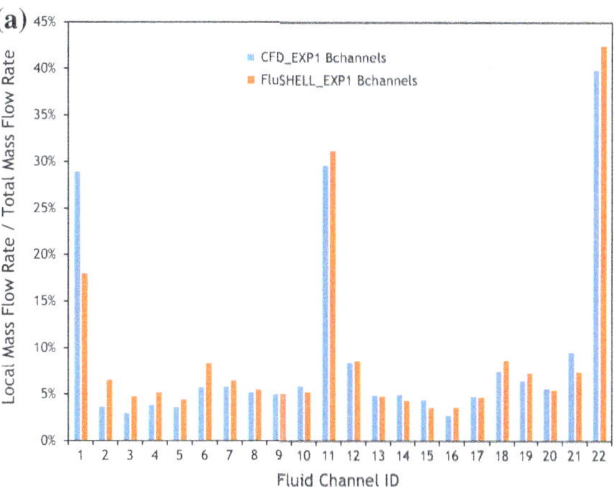

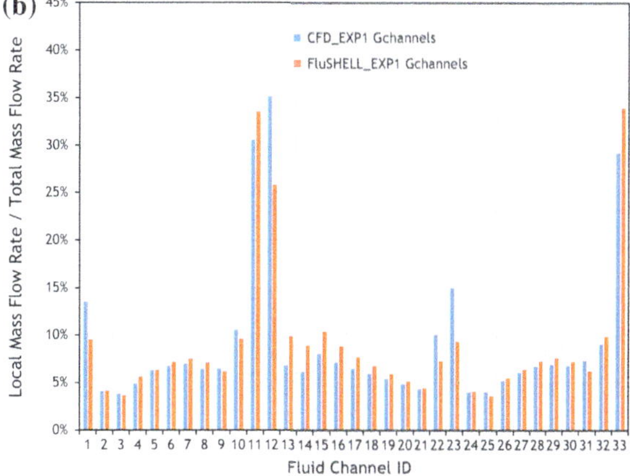

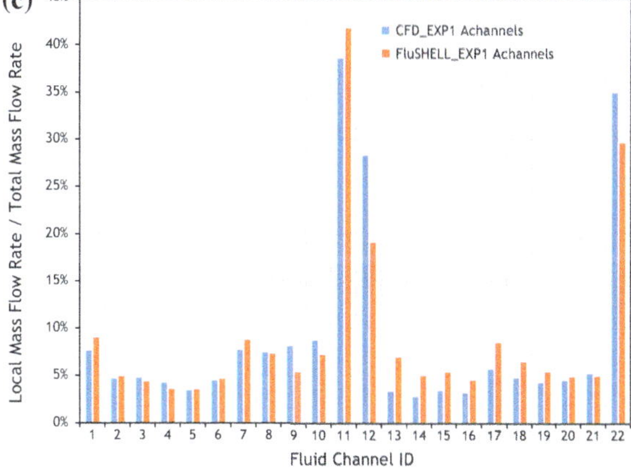

Table 5.10 Mass flow rate deviations between FluSHELL and CFD

Mass flow rate deviations— $\left|\frac{q_{channel}^{FluSHELL}-q_{channel}^{CFD}}{q_{channel}^{CFD}}\right| \cdot 100$								
EXP1 (%)	EXP2 (%)	EXP3 (%)	EXP4 (%)	EXP5 (%)	EXP6 (%)	EXP7 (%)	EXP8 (%)	EXP9 (%)
Bchannels 20	22	23	20	20	21	20	20	20
Gchannels 14	13	12	16	14	13	19	16	14
Achannels 27	26	26	27	26	26	26	26	26
Average **19**	**19**	**19**	**20**	**19**	**19**	**21**	**20**	**19**

Table 5.11 Pressure drops predicted using CFD and FluSHELL. Relative deviations

Pressure drop, ΔP [Pa]

Sim. ID	CFD	FluSHELL	FluSHELL-CFD (%)
EXP1	3445	4023	+17
EXP2	4045	5198	+29
EXP3	6140	8613	+40
EXP4	5909	6431	+9
EXP5	6905	8232	+19
EXP6	9551	12011	+26
EXP7	7443	7436	+0
EXP8	8905	9715	+9
EXP9	11305	13381	+18
Average	NA	NA	+19

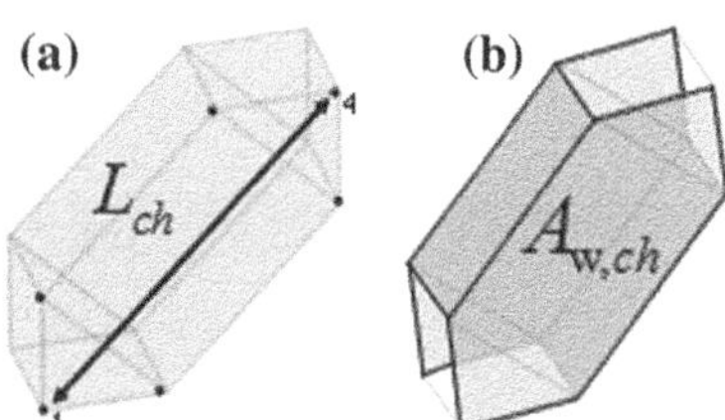

Fig. 5.24 Geometrical attributes of the fluid channels as considered in the FluSHELL tool

5.4 Conclusions

In this chapter, the FLUSHELL performance has been directly compared with experiments. FluSHELL considers the coil/washer system as being perfectly adiabatic and this comparison evidenced that the non-adiabatic conditions of the actual experimental setup promote significant temperature over predictions.

In a commercial shell-type transformer it is not yet fully understood if the coil/washer system is effectively adiabatic or not, and this has not been addressed in this work. Instead, a new adiabatic three-dimensional CFD model has been developed to enable a rigorous conceptual validation of the FluSHELL tool.

In consequence of a comprehensive comparison between FluSHELL and this new CFD model it has been observed that:

1. FluSHELL has been able to compute coherent mass and energy balances, with low residuals and within an average simulation time of 2.2 min which is two orders of magnitude lower than CFD that in turn exhibit an average simulation time of 262 min.
2. FluSHELL revealed an average accuracy of 1.8 °C in the average temperature of the turns and 2.4 °C in the hottest temperature. This occurred consistently for all the experiments and within range of heat fluxes and oil velocities (11-26 W. dm^{-2} and 10.9–19 $cm.s^{-1}$, respectively) that represents common design practices.
3. FluSHELL has been able to identify the number of the turn where the hottest temperature occurs as well as where that temperature occurs within the turn. The hottest temperature occurred consistently either in the innermost or in the outermost turns which are located beneath the respective insulation frames.
4. FluSHELL has been able to reproduce quite well the oil temperature pattern evaluated over 77 control surfaces that were created over the whole fluid domain. FluSHELL predicted the oil temperatures inside the coil with an average deviation of 0.5 °C.
5. FluSHELL predictions agree quite well with the oil flow pattern inside the coil over the same 77 control surfaces. In average, the mass flow rate distribution over these surfaces has been predicted with a deviation of 20% comparing with CFD.
6. The pressure drop predicted between the inlet and the outlet of the coil is approximately 20% higher in FluSHELL than in CFD.

Based on these results, and considering this is a novel application of this concept to shell-type coils, the FluSHELL tool is considered conceptually validated.

Chapter 6
Conclusions and Future Work

This work involved four core activities:

- the design and manufacturing of a scaled down experimental setup in order to validate the numerical approaches used/developed in the remaining activities of the work—Chap. 2;
- the application of a commercial CFD code both to model the experimental setup and to correlate transport properties (e.g. friction and heat transfer coefficients) —Chaps. 3 and 4;
- the development of a novel thermal-hydraulic network model which uses the transport properties correlated using CFD—Chap. 4;
- the validation of both numerical approaches against experiments—Chaps. 3 and 5.

The experimental setup has been successfully designed, manufactured and used to validate the CFD which has been observed to be a valid technique for this purpose. Afterwards, a thermal-hydraulic network tool has been developed and conceptually validated both against experiments and CFD.

The main advantage of this proprietary tool is the shorter time-to-solution and accurate predictions when comparing with CFD.

Section 6.1 compiles the main conclusions about these activities and discusses the potential relevance of the FluSHELL tool for a transformer designer. The characteristics and outputs of this novel tool are put into perspective against the current design practices.

Section 6.2 proposes future related activities.

© Springer International Publishing AG 2018
H. Campelo, *FluSHELL – A Tool for Thermal Modelling and Simulation of Windings for Large Shell-Type Power Transformers*, Springer Theses,
https://doi.org/10.1007/978-3-319-72703-5_6

6.1 Conclusions

Power transformers, either core-type or shell-type, are static electrical machines with two main components: the windings and the magnetic core. These two components together are responsible for the voltage transformation which is the basic need underpinning the ubiquitous use of these machines over the electrical networks worldwide. This voltage transformation is accomplished with high thermodynamic efficiencies (>99.5%), although the magnitude of the residual energy lost is on the order of hundreds of kW. Most of this residual energy is dissipated in the windings under the form of heat due to resistive and inductive effects when high electrical currents are circulated through its metallic electrical conductors.

Thus, at the design stage, the windings are one of the upmost concerns and its thermal performance is one of the key aspects that need to be addressed by the design engineers. At that stage, and from this thermal perspective, the design engineers must guarantee that the windings and the circulating cooling fluid, under steady-state conditions, will not exceed certain temperature levels that will be further subject of guarantee. After manufacturing and before the commissioning of the transformer, the machines are tested and the guaranteed temperature levels must not be exceeded otherwise the transformer may not be finally accepted or maybe de-rated with subsequent impact on its market price.

As a result, from the perspective of the design engineer it is of upmost importance to have a reliable thermal tool that enables, at least, the calculation of the temperature levels guaranteed: the maximum fluid temperature, the average temperature of the windings and the maximum temperature of the windings.

If the context demands a standardized product requiring less customization and the operational margins of the business are regular, the drivers for more complex tools are attenuated. Although, due to the strong de-regulation in the energy markets over the last years, the power transformer industry is under strong market pressure to improve operational margins while maintaining competitive prices.

This context demands differentiation and innovative solutions, thus the detail of the tools available for design engineers assumes special relevance and it is important to provide means for evaluating changes due to variations in a wider range specific design parameters.

As discussed in a recent working group from the International Council on Large Electric Systems (CIGRE) the thermal models currently used to design shell-type windings are simpler. The generalized current design practice comprises the use of simpler unidimensional thermal expressions to predict exclusively the guaranteed temperature levels (Cigre 2016). In turn the current design practices for core-type windings already involve the use of thermal-hydraulic networks models. Under some circumstances, and for both transformer technologies, the use of Computational Fluid Dynamics (CFD) has also been reported. However, CFD is currently regarded as virtual laboratory used mainly for R&D purposes and not as a direct design tool.

On the other hand, thermal-hydraulic network tools are proven to be effective thermal modelling approaches that are being applied to core-type transformers since the 1980s, when the first known model of this kind has been reported by Oliver (1980). The relevance of these models to power transformer industry (either manufacturers or users) derives from the enhanced sensitivity to a wider range of parameters and the detailed information that can be computed in few minutes. This time-to-solution is acknowledged to be a crucial advantage comparing with CFD, namely because it allows a faster integration with the other proprietary tools that are already part of the design-cycle. So, at some extent, the thermal-hydraulic network tools represent also a certain independence from external commercial tools.

In a shell-type transformer the windings are composed by groups of coils which are separated by washers and each having identical geometries. Thus, from a thermal-hydraulic perspective these coils are independent. Each coil combined with its adjacent components (spacers, washers and insulation frames) form a single coil/washer system and this system comprises the focus of this work.

In this work, a thermal-hydraulic network tool (FluSHELL) has been first developed to solve this type of coil/washer systems.

FluSHELL discretizes all the fluid and solid domains into fluid channels and turn segments. The accuracy of this tool, depends on the use of adequate correlations for friction and heat transfer coefficients. Thus, these coefficients have been extracted from CFD, correlated and implemented in FluSHELL.

This tool provides means for a designer to compute:

- the temperatures in each fluid channel and turn segment enabling a proper quantification and localization of the hottest temperature;
- the detailed mass flow distribution in each fluid channel enabling a better management of the potential streaming electrification induce by localized high fluid velocities;
- the pressure drop corresponding to a target average fluid velocity or the total mass flow rate corresponding to a target fluid pressure. While designing the whole hydraulic circuit this information might be useful to choose adequate pumps;

Moreover, an experimental setup has been built to validate both CFD and FluSHELL results. The CFD temperature predictions have now been validated complementing the isothermal CFD flow patterns previously validated in 2007 by Gomes et al. (2007b).

When comparing FluSHELL results with both experiments and CFD it has been observed that:

- the relevant quantities are in good agreement both in terms of magnitude and spatial distribution.
- the average time-to-solution is two orders of magnitude lower than CFD (~ 2.2. min).

FluSHELL is now expected to promote improved design decisions as it provide means for the designer to change:

- the losses magnitude and its distribution between the turns;
- the size of the inlet and the outlet region;
- the number and size of the spacers;
- shape and size of the inner and outer insulation frames.

Complementary, this is a transformer technology with fewer units worldwide and the public information is not as dense and widespread as to what concerns with core-type transformers. In this sense, this work disseminates the most relevant heat transfer mechanisms influencing the thermal performance of this type of transformers, thus representing a significant step towards a wider awareness about this technology.

6.2 Future Work

The main goal of this work was providing means for an accurate prediction of the maximum and average temperatures occurring in different coil/washer systems inside a shell-type transformer to finally promote better design decisions. At this stage the means have been provided with the FluSHELL tool and it has been conceptually validated.

In terms of further CFD modelling and Experiments this work raised questions, motivated other challenges and showed that it is still relevant to:

- conduct more experiments under ON conditions. The experimental setup built showed a reproducible steady-state performance for a range of common OD cooling modes. Despite of this, there are some shell-type units in the market operating under ON cooling modes. As these machines are heavy rated and mainly installed at Transmission levels, there is no specific pressure for dimensions and therefore the operation under ON is interesting from a maintenance perspective (e.g. no pumps to maintain). The experimental setup built permits circulating the cooling fluid through a closed loop that by-passes the gear pump, thus further activities under buoyancy driven flows associated with ON cooling modes are understood to be feasible and relevant;
- conduct more experiments with other cooling fluids. The natural esters seem to be in this moment assuming higher political relevance due to its lower flammability and biodegradability comparing with the current mainstream cooling fluids which are the naphthenic mineral oils. The experimental setup built might be used to assess the comparative thermal performance of natural esters against other fluids;
- better understand the fluid flow in the peripheries of the coil/washer system. It seems reasonable to assume that adjacent coils do not exchange mass or heat between each other. The pressboard washers separating the coils and the similar

cooling conditions around each single coil seems to guarantee that. Although, it has been observed that the maximum temperature consistently occurs in the turns located beneath the insulation frames and it has been also observed that the magnitude of that temperature is significantly influenced by the heat transfer conditions of the neighbourhood, namely when the CFD of the Scale Model has been compared against an Adiabatic CFD Model. As mentioned in Chap. 1, there are some additional fluid flow paths around the coil/washer system to remove additional heat being generated namely in the Magnetic Circuit and the T-Beams. Therefore, the CFD evaluation of the interaction between these flow paths and the insulation frames is relevant to identify whether the coil/washer system is effectively adiabatic or not;

- better understand the thermal performance of different insulation frames. So far, the geometry of these frames has been simplified. In shell-type, contrarily to core-type, the computational domain cannot be easily reduced without losing relevant information. This has been initiated during this work but due to time constraints this task has been abandoned.

  However, these frames can have different geometries and might comprise additional fluid flow paths to reduce the local temperatures. In practice, it is also quite common the use of metallic static rings next to the innermost and outermost turns. These details need to be parametrically modelled in CFD as they seem to have a significant impact on the local maximum temperatures observed inside each winding (the commonly called Hot-Spot Factors).

  It is noteworthy that, in this work, two types of fluid channels have been identified and implemented: the radial and the transverse channels. These numerical activities more focused near the insulation frames might enable the identification of fluid channels with different characteristics.

- better understand the fluid flow in tank. The fluid before entering the coils suffers a sudden infinite expansion which is assumed to create uniform pressure conditions at the entrance of the coils. The literature so far seems to reinforce this idea (Cigre 2016), although the evidences are limited and further CFD simulations are needed to better quantify this. It is noteworthy that an unbalanced flow distribution between coils or between phases may influence the efficiency of the thermal performance of the transformer;

- better understand the fluid flow in the bottom and top insulation structures. In practice the bottom and top part of the coils are protected with additional cellulosic structures that have been neglected in this work. In principle, these structures only impose an additional head loss to the fluid flow that need to be taken into account at the design stage. Although the magnitude of this head loss and its impact on the flow distribution inside the coil needs to clearly assessed through adequately parameterized CFD simulations;

- evaluate other geometrical arrangements of the coil and of the spacers. In the most common coil arrangement, which has been studied in this work, the coil comprises two independent cooling sides and the interfaces between two consecutive turns are guaranteed by pressboard strips that do not allow any direct contact with oil. In this technology, contrarily to core-type, the copper conductors are winded with their highest dimension in the vertical position which is then blocked from the direct contact with the oil due to the neighbouring turns. However, there is a significant amount of potential heat transfer area between the turns that is not being efficiently used. The lack of public information about this technology does not enable a clear assessment whether this is already being practiced or not.

In this work, it has been observed that an accidental exposure of that dimension to oil has promoted significant lower average temperatures. This seems to indicate that there is room to improve the heat transfer efficiency of this coil arrangement (e.g. more compact transformers or transformers with an extended lifetime). In this way, further numerical analysis of different arrangements and validation in this experimental setup are understood to be relevant.

All these activities will require further transfer of knowledge to the FluSHELL tool which is the end-tool to be delivered to a transformer designer. Perhaps, the most significant evolution is the further adaptation of the hydraulic and heat transfer models to unsteady conditions.

This evolution to unsteady conditions, which are the operating conditions in the electrical networks, widens the application scope of this tool enabling the prediction of unsteady temperatures but also enabling a precise prediction of how the weakest components of the transformer degrade during its life-cycle. This evolution coupled with adequate ageing mechanisms is understood to be a medium term attractive, both for manufacturers and customers, and one of the main added values of the FluSHELL tool.

References

Cigre. (2016). *WG A2.38 brochure, transformers thermal modelling*. Draft Version 5.

Gomes, P. J., Sousa, R. G., Dias, M. M., & Lopes, J. C. B. (2007b). Large power transformer cooling—flow simulation and PIV analysis in an experimental prototype. In *Advanced Research Workshop on Transformers, Baiona* (pp. 113–129).

Oliver, A. J. (1980). Estimation of transformer winding temperatures and coolant flows using a general network method. In *IEEE Proceedings C—Generation, Transmission and Distribution IET* (pp. 395–405). https://doi.org/10.1049/ip-c:19800061.

The manufacturer's authorised representative in the EU is Springer
Nature Customer Service Centre GmbH, Europaplatz 3, 69115 Heidelberg,
Germany. If you have any concerns regarding our products, please
contact ProductSafety@springernature.com

Printed and bound by CPI Group (UK) Ltd, Croydon, CR0 4YY

28/11/2025

02007692-0008